MY VIEW FROM THE SILVER TAHOE

by
Dorothy "Dotty" Sanders

Blessings and peace,
Dotty Sanders

ISBN No. 0-9772627-1-5

This book is dedicated to the

memory of

my mother, "Mama Dor,"

and

in honor of the many

prayer warriors from

"sea to shining sea."

Acknowledgements

Appreciation is expressed to:

The Kentucky United Methodist Homes
for Children and Youth

Melbourne United Methodist Church

Robke Chevrolet (furnished the Tahoe for the trip)

All contributors and supporters

Friends

Family

Introduction

This book is a day by day account (journal) of "dream fulfillment".

For over forty years my husband, Dewey, had a dream to walk across the United States and on February 10, 2004 he took the first steps toward making this dream a reality. His journey began at First Landing-Virginia Beach, Virginia and proceeded for one hundred forty-six days — ending in San Francisco, CA by walking across the Golden Gate Bridge (GGB) on July 4, 2004 accompanied by an entourage of 38 friends.

From the Atlantic Ocean to the Pacific Ocean we traveled the backroads of mid-America through 14 states and crossing many rivers along the way. Each day was a paradox — alike yet different. Dewey walked and I drove, BUT God intervened each day by giving us unique experiences, new friends and innumerable blessings.

My prayers for safety, good health and good weather were answered in abundance as God kept us safe and healthy and gave us unbelievable weather — considering we traveled in winter, spring and summer! Would you believe we had only 5 days of wash-out rain and 2 days of measurable snow (one on June 9th in Austin, NV)? God truly parted the Red Sea and gave us dry land on which to walk and drive.

Throughout this incredible journey the "gift of friends" was a special blessing. Their phone calls, daily notes, surprise visits and prayers enabled us to stay focused and continue trekking. The success of the venture was possible because people cared about us and our cause.

Dewey walked 3,355 miles and I drove 17,500 miles "Coast to Coast for Kids." We raised money for the SHIPS and SIGNALS Program written by Lane Stonecypher on staff at the KY United Methodist Homes for Children and Youth at the Versailles, KY campus. We are indebted to the many people who have already contributed to the Walk.

This journal was transcribed as I wrote it during the journey with all the abbreviations, expressions, personal shorthand, and occasional mistakes used in my daily writing. To make it as authentic as possible, I edited only when it was necessary for the meaning of the material. Through this, I hope readers will experience the trip from my perspective in a first-hand way.

All profits from this printing of the book will benefit the SHIPS and SIGNALS program.

The Journey Begins

February 8, 2004

It was a day to remember! Wonderful church service with Dr. Hahn (Methodist Home CEO) speaking — used Joseph's Dream(s) as his text — followed by communion and "hands-on" prayer for our safety and good health. At 3:00PM we gathered at Alexandria UMC for the kick-off dinner with 180+ in attendance including church, neighbors, co-workers and family. Speakers included Dr. Hahn, Lane Stonecypher, DS Terry Reffett, Jeana Walter, Dewey and our precious son, Craig. Wes Daniels was MC and Sue Daniels gave the invocation. The committee prepared a delicious meal (ham, green beans, scalloped potatoes, slaw, hot rolls and a variety of desserts). We enjoyed the day with family and friends. It was a special memory to treasure. We left Alexandria at 6:20PM and traveled to Charleston, WV. We arrived there around 10:30 to 11:00PM. Jingle enjoyed ham from the dinner and slept all night with her daddy Dewey.

Weather: great, clear & cold

February 9, Monday

We left Charleston, WV at 6:30AM — stopping for coffee and biscuits at McDonalds before getting on the toll road. We arrived in Suffolk, VA at 3:00PM and checked into Holiday Inn. Jingle couldn't stay in motel tonight, pouting about it! We ate dinner at George and Steve's Steakhouse — old fashioned eating. Early to bed as we have an early morning — the walk begins in AM — (Dewey met with a police officer to advise him re walk).

Weather: AM clear and mild
PM rainy and cool

February 10, Tuesday-Day 1, Walk begins *Miles walked: 26.21*

We arose at 5:30AM to go to VA Beach (First Landing) to get started. The video team from MH was there along with Jeana. We ate a quick breakfast at McDonald's because the Pancake House we had planned to eat at was closed on Tuesday. Dewey didn't get to start walking until around 11:00AM and he had to go "around the barn" to avoid Interstate bridges to get back to Suffolk. He walked 26.21 miles today (until 5:00PM). He is tired and so are Jingle Bell and I. But the venture has begun.

Weather: no rain today
Temp: high of 52 degrees

February 11, Wednesday-Day 2 *Miles walked 27:08*

We got up at 6:00AM to leave by 7:00. We had to return to VA Beach and pick up where the walk ended the day before. Dewey walked 27.08 miles today. The highlight today was being escorted across a drawbridge by police as it was a non-pedestrian bridge but safer route according to police Lt. Joseph Patterson. We ended today on Rt. 125 by a lovely Methodist Church. We will return to that point tomorrow for the continuing journey. We are staying in the Quality Inn in Petersburg for 2 or 3 nights so this will give us a break from in and out sleeping! Ann Crowe will join us tomorrow for several days. She will walk some with Dewey and ride some with me. I have been sticking fairly close to Dewey since the route has been "fuzzy". Jingle Bell and I are cheering Dewey on and Jingle has even walked some! We are truly blessed.

PS: Ate at Virginia Diner in Wakefield. Its motto is "a Legend in a nutshell". Claims to be peanut capital of the world since 1929.

Weather: no rain today
Temp: high of 50 degrees.

February 12, Thursday-Day 3 *Miles walked: 27.55*

Today was the first day of inclement weather. We had to return to the Methodist Church on 125 to begin the walk. Dewey had to walk in rain **all day**. On about mile 2 he turned his ankle dodging a truck. So he limped all day, but **never quit**. We ate at the Virginia Diner in Wakefield, VA which is a world renowned eating establishment. Food is great and atmosphere quaint. Ann Crowe arrived by train to join us for a few days. She will walk some with Dewey and help me drive too. Her train was an hour late from Washington DC. Oh, the perils of railroading. I did laundry for the first time! Dewey's clothes from today were nice and soggy! We're getting to bed around 10:00PM and arising at 6:00AM. We ate breakfast in the motel (Quality Inn) and headed down the road. We have to return to Route 620 in AM (3 miles to go on this road).

Weather: rainy
Temp: in upper 30's most of day

February 13, Friday-Day 4 *Miles walked: 22.5*

FRIDAY THE THIRTEENTH

We got up at 6:00AM, ate breakfast in the motel and drove back to Rt. 620. On the way we were pulled over by police car for exceeding speed limit — officer was kind — only gave us a warning (I **wasn't** driving). Ann walked 3+ miles with Dewey in the AM and then I picked her up and Dewey continued walking for a total of 10+ miles. He joined us at the Virginia Diner for lunch. Ann walked with him again in the PM for 3+ more miles. Dewey walked a total of 22.5 miles today — total of 103+ so far. We ate one last time at the Virginia Diner. We are going to spend tonight and Saturday here in Petersburg at the **Quality Inn**. It is nice not having to move everyday. Having Ann here with us is a blessing. She walked some, rode some and drove some. God is taking wonderful care of us — We are truly grateful.

The route today was highway 620 for 3 miles and rest on 460.

Weather: fair
Temp: in the mid to upper 40's most of day

February 14, Saturday-Day 5 *Miles walked: 23.14*

HAPPY VALENTINE'S DAY

We got up at 6:15AM, dressed, ate breakfast at the motel and drove E on 460 to the beginning point for the day. Ann walked with Dewey for 3.7 miles and will walk again in the PM. We lunched at McDonald's (not the VA Diner for sure). Dewey walked another 5 hours. He walked a total of 23.14 miles today. His total so far is 126+ miles. Ann walked 7 miles for a total so far of 13 miles. I have a total of 0 miles walking but many, many, miles driving. Jingle couldn't walk the last 2 days because the roads haven't been safe. Dewey has had to dodge trucks and cars. Highway 1 is a US highway, but little shoulder to walk on. We'll move on tomorrow to Richmond and stay for 2 nights. Ann and I shopped at the Dollar Tree, but since we have very little room in the car, I refrained from any purchases except batteries for Dewey's camera. Blessings continue.

Weather: cloudy
Temp: high of 54 degrees

February 15, Sunday-Day 6 *Miles walked: 9.64*

We left a week ago today. We attended an 8:30AM service at **Ivey**

Memorial United Methodist Church in Colonial Heights, VA. We had brunch at O'Charley's and Dewey began his walk. Ann and I found a room for tonight at Red Roof in S. Richmond. Jingle can stay with us. Ann walked 3+ miles with Dewey in latter part of afternoon. We will go out for a bite of dinner at the Olive Garden Restaurant. I called a college friend in Ashland, VA and we plan to meet tomorrow (2-16) and visit a spell. We share a lot of MWC memories. It should be a special time for both of us. Dewey walked 9.64 miles today. He has now walked a total of 136+ miles. God is our daily caretaker.

Weather: cloudy
Temp: high of 41 degrees

February 16, Monday-Day 7 *Miles walked: 22.12*

We arose at 6:30AM and Dewey began walking around 8:30AM. Ann and I drove on to Fredericksburg and got a room for the next two nights. We returned south via 95 to meet Dewey for lunch. We ate at a café that is part of a grocery store called Ukrop's. The food was great — fixed our own salads/soup etc. Ann and I had left our laundry to be done at a laundromat. We picked it up after lunch and Ann walked for a while this afternoon. I met a classmate from MWC in Ashland. Felicia and I were suite mates during our freshman year and roommates our sophomore year. It was great visiting with her for a while. Dewey walked until 5:00PM today for 22.12 miles. He's now walked 158+ miles. I've driven many more. We ate supper at Aunt Sarah's Pancake House-food was good but service was poor. Jingle enjoyed her chicken from the dinner. God is giving us many memories.

Weather: sunny
Temp: 30's

February 17, Tuesday-Day 8 *Miles walked: 22.32*

We got up at 6:45AM, dressed and ate breakfast at McDonald's. Had to drive Dewey back almost 30 miles to begin his walk today. Ann walked 5 miles with him while I waited and shopped at UKrop's for a few needed items. We lunched at Subway. Dewey and Ann resumed walking after lunch. I walked Jingle and read awhile until Ann was ready to be picked up. She and I returned to Fredericksburg and she rested some. I checked out the rt thru Fredericksburg and the spot where Dewey will pick up Rt. 17 which goes to Winchester, VA. Dewey walked a total of 22+ miles today for a grand total of 182+ miles. There were snow flurries off and on today, but nothing serious. We ate dinner tonight at the Cracker Barrel. God continues to bless us with good weather and is keeping us healthy.

Weather: cloudy
Temp: 30's

February 18, Wednesday-Day 9 *Miles walked: 23.69*

Today began very early as Dewey's phone rang at 5:40AM. We knew it must be an emergency! It was. Ann's dog, Ivory, had died. What a shock! Dewey had already planned to walk only a half day today and he's glad. Ann leaves us today for a meeting in DC area. We will drive her to Alexandria and she will attend her meeting and fly home on Friday evening. We took Ann to the Sheraton in Alexandria and returned around 4:00. We did laundry and then went to the Forum Family Restaurant to eat. The food was "home cooking" delicious. It offered Italian, Greek and American cuisine. Dewey had country fried steak, baked pot., soup, salad, carrots, dessert and home made bread. I had crab cakes, salad, green beans, baked potato and bread…all for $8.95 a piece — Yummy. Jingle liked her bites too. We leave in the AM for Winchester, VA. Thank you God for your continued watchcare and comfort in times of need.

Weather: sunny
Temp: 40's

February 19, Thursday-Day 10 *Miles walked: 27.08*

We left Fredericksburg around 8:15AM after eating a quick breakfast at McDonalds. Dewey began walking on highway 17 towards Winchester, VA. It will take approximately 3 days because it is 80+ miles from Fredericksburg. I drove ahead and made reservations at a Hampton Inn in Warrenton, VA for Thurs. & Fri. night. Maybe Sat. as well; depends on progress. I picked up lunch at M&P Restaurant and took it back to Dewey. I returned to restaurant and waited for Dewey to make it to this point. He walked a total of 27.08 miles today. The Hampton Inns are always nice. Our room is on 1st floor in the back which makes walking Jingle Bell easier. We ate dinner at "Country Cookin" restaurant. It was fair, but I was able to get vegetables. (Sort of like a Golden Corral, but fewer menu items). Going to bed early — blessings continue.

Weather: sunny
Temp: high in the mid 50's

February 20, Friday-Day11 *Miles walked: 25.39*

We got up around 7:14AM and had breakfast at the motel. Hampton Inns always serve **great** breakfasts! I drove Dewey back to the place he stopped yesterday. I shopped a little and found out about Dewey's Rx at Rite Aid. Dewey walked a total of 25.39 miles today. I picked him up for lunch at 1:00 and he returned to walk at 2:00. He walked until 4:30. We picked up Rx and returned to motel. We got ready and went out to dinner at Molly's Irish Pub. The food was good and atmosphere quaint but loud. We returned to motel and Jingle Bell went to bed immediately and is already fast asleep! We plan to leave Warrenton in AM and head for Winchester, VA. God continues to watch over us.

Weather: sunny
Temp: high in low 60's

February 21, Saturday-Day 12 *Miles walked: 18.63*

We got up at 6:30AM and ate breakfast in motel (Hampton), packed the car to move on to Winchester, VA. I took Dewey to place he stopped on Friday; however, we first checked out the route as 17 became interstate for 5 miles so he couldn't go that way. He followed Rt. 110 to 112 to 50. When he got to 50, there were wind gusts 30 to 40 miles an hour. This really affected his ability to make much time. While he was walking, I drove on to Winchester and got a room for 2 nights at **Red Roof**. Jingle is allowed to stay and we got a 1st floor room with microwave and refrigerator and you can park by the room. We like exterior corridors with Jingle Bell along. Dewey walked 18.63 miles today and he's bushed after fighting the wind all day. He's now walked 275+ miles. We ate dinner at Hoss' (like a Ponderosa). Jingle liked her chicken treat! We are doing laundry now, then to bed. Thank you God for everything.

Weather: partly cloudy, wind a real factor today
Temp: high in 50's and ended in 40's

February 22, Sunday-Day 13 *Miles walked: 10.03*

We got up at 6:45AM as we wanted to attend an 8:30 worship service. We went to 1st Methodist, in downtown Winchester, VA. It was an uplifting service in which the minister spoke on Mt. top experiences. (Scripture+Transfiguration). A layman gave scripture reading from memory (told the story). I requested prayer for Dewey's walk and a lady gave me a $10.00 check for the cause. It touched me. I will write her a note of thanks. We received 2 mugs from church as 1st time visitors. I ate some sugar-free ice cream at the mall while Dewey walked from noon to 3:00. He has walked a total of 285+ miles to date. We ate an early dinner at an Italian restaurant. Food was good and tasted authentic. Dewey is itching pretty badly so he got some Benedryl to try and stop it. I'm doing a load of laundry. It seems I never get caught up! We plan to go to bed early and Dewey hopes to get through Winchester and start through the Mts in WV also tomorrow. Weather is supposed to break tomorrow, but possibility of snow flurries or showers on Tuesday-Thursday. I pray God will continue to bless us with good weather.

Weather: sunny & windy
Temp: 40's

February 23, Monday-Day 14 *Miles walked: 21.83*

Dewey had another restless night last night — itching back and restless leg syndrome. Craig called his dad to check on him. He called in an RX for him and I had to seek out the pharmacy to pick up the RX. Dewey walked 21.83 miles today on Highway 50 towards WVA. He reached the WVA line late this afternoon. He may take tomorrow off and rest and regroup. Snow is predicted for tomorrow so we'll see! Jingle Bell still hasn't been able to do any walking-too many trucks. I met Dewey at lunchtime today and we had a sandwich together. He continued walking until 4:30. Tonight we ate at Country Buffet — but it wasn't too good. Jingle liked her roast beef, however. We're staying another night at Red Roof Inn in Winchester, VA. If Dewey takes tomorrow off from walking, we will stay one more night here. Our blessings are truly from "On High."

Weather: sunny
Temp: 40's & low 50's

February 24, Tuesday-Day 15 *Miles walked: off day*

Dewey is taking today off from walking. Jim McCord encouraged him to do this from time to time to keep it all in perspective. I spent time in a Mall while Dewey rested. We both napped in the afternoon along with

Jingle Bell. We ate lunch at Cracker Barrel and dinner at Bob Evans. It has been an uneventful day, but given Dewey some time to rest. We will leave here tomorrow and wind along Route 50 in WVA. We aren't sure what is available for eating or sleeping but I will seek it out while Dewey is walking. It was a rainy/snowy (no accumulation) day today, but tomorrow promises to be better. We are grateful for prayers being lifted up for us each day. We feel every one of them.

Weather: rain/snow
Temp: 30's

February 25, Wednesday-Day 16 *Miles walked: 21.38*

This was a day spent in the car. I must have driven 200 miles back and forth etc. I had to go 35+ miles ahead to find a motel that would accept doggies. It's what our friends the McCanns call "Spacious", a private joke of ours. In other words, it is not a "Holiday Inn." But Jingle gets to stay with us! Dewey walked 21.38 miles today. He has now walked a total of 327+ miles. He is walking on Hwy 50 right thru the Mts. and some of the Mts. are high and curvy. We ate at a truck stop restaurant tonight in Burlington, WVA. There is a Methodist Children's Home in Burlington right on hywy 50. The weather is supposed to be nice for several days. We hope to be out of Mts. by then. We were in a huge backup on hwy 50 near Romney, WVA. There was a wreck and fire truck, police and life squad was in abundance everywhere. Thank you God for safety.

Weather: sunny
Temp: 30's and low 40's

February 26, Thursday-Day 17 *Miles walked: 20*

Today began around 7:00AM. We got up and drove down the Mt. and ate breakfast. We then drove Dewey into town to begin his walk. However, before he started he went to the Dr. because the itching and rash have been really bad for him. The Dr. and his staff were really interested in Dewey's trek. Dewey doesn't feel much better tonight. Hopefully, by morning the meds will begin to make a difference. Dewey started walking around 11:00AM and walked until 4:30. We ate supper at same place we did last night-not many choices here in "the outback". I got my hair done today and it makes me feel better. Jingle enjoyed some treats of chicken today. Dewey will begin walking up really high and curvy mts. tomorrow — 19 miles of what to me would be sheer torture! The phones aren't working very well here in Mt. Storm, W Va — too high up in the clouds. I was waiting for Dewey at one point today and a State Trooper checked to see if everything was okay. I was appreciative. Dewey walked 20 miles.

Weather: fair & sunny
Temp: 40's for high and lows in 20's

February 27, Friday-Day 18 *Miles walked: 20.12*

We got up a little later today as Dewey didn't get much sleep — still itching. We ate breakfast at Van Myra Inn where we ate breakfast & dinner yesterday. We ate dinner tonight at the Rt. 50 Café. Good home cooked food. Dewey walked straight up for much of today. He walked 20.12 miles. We will leave Mt. Storm in the AM and I'll probably drive 50+ miles to Bridgeport, WV where we'll stay for 3 or 4 days. I'll then back track and pick Dewey up in the PM. I tracked the rt. part of the way today. Tomorrow Dewey will walk the 7 miles through MD and continue then on 50 in WVA. The weather has been nice. A couple told me today that the rt Dewey's walking had 2 feet of snow 2 weeks ago. Thank you God for the blessing of sunshine and milder temps. You, God, have been the constant in this venture.

Weather: fair & sunny
Temp: lows in 30's and high in 40's

February 28, Saturday-Day 19 *Miles walked: 21.18*

Today began at 6:45AM at Mountaineer Motel at Mt. Storm, WV. We got ready and went to Rt 50 Café for breakfast. Many locals were there having coffee. We returned to the motel to pack the car and head for our next destination — Bridgeport, WV. I will be coming and going for the next few days taking & retrieving Dewey to and from walking destinations. He walked 21.18 miles today for a grand total of 388+ miles. I found a nice motel — Holiday Inn in Bridgeport and with our coupon it is $45.00 a night and Jingle gets to stay with us. I found a great restaurant in Grafton, WV the home of Mother's Day founder, Anna Jarvis. The restaurant is called 4 Corners. The hostess gave Dewey his dinner — her contribution to the walk. We had a great surprise tonight. Tim and Linda McCann came to spend the night and give Dewey support for the walk. They are great and caring friends. We are truly blessed with so many who care about this adventure. Thank you God for your continued guidance and watchcare.

Weather: beautiful and mild
Temp: high in 50's

February 29, Sunday-Day 20 *Miles walked: off day*

SADIE HAWKINS DAY

We arose early and went to church at Bridgeport United Methodist Church in Bridgeport, WV. to their 8:30 service. It is a large, beautiful church with a gorgeous stained glass window. It reminded us a little of the Old Centenary UMC in Lexington. There were 200 or more in attendance at this early hour. The church building is 50 years old and was given by a wealthy man that many years ago. We came back after church and the McCanns and we went to breakfast at our favorite restaurant in Grafton. Food excellent! We went to see the Mother's Day Memorial (Anna Jarvis) but it is being refurbished — scheduled to reopen in April. Dewey did not walk today because of a sore foot. He will resume the trek tomorrow. We rested some and I went to a laundromat to do 3 loads of clothes. Everyone in the motel seems to know Jingle Bell. Most call her by name. We ate dinner tonight at an Italian eatery; Minard's famous for Italian food since 1937. Our blessings are legion. We are truly grateful.

Weather: beautiful and mild
Temp: high in low 60's

March 1, Monday-Day 21 *Miles walked: 24.14*

We got up early and drove to Grafton and ate breakfast at 4 Corners. I then drove Dewey to his starting spot. I went to Wal-Mart and got a few fruit items for Dewey and me. I wrote cards for a while and read a bit. Dewey walked 11 miles this AM and then I picked him up and we ate lunch at 4 Corners. We ate their $4.95 buffet which included 2 kinds of soup, great salad bar, sandwich bar (turkey, ham, roast beef, cheeses) hot bar including chicken and dumplings, goulash, fried chicken, corn, mashed pot, mac & cheese. When we left Dewey bought doughnuts, coconut cream pie and peanut butter pie to bring home — all delicious. Dewey walked 24.14 miles. Dewey's total mileage is 412+ miles. His feet are sore again. Hopefully they'll feel better by morning. We ate in room tonight. I got a salad and some fried cheese sticks and Dewey ate fruit and desserts. Dewey has talked with several people today and had e-mails from others. Really lifts his spirits to hear from home. Dewey will walk towards Clarksburg & Parkersburg tomorrow. Dewey avoided being hit by a car today! God is definitely his co-walker.

Weather: fair & mild, partly cloudy in PM
Temp: reached 60's today

March 2, Tuesday-Day 22 *Miles walked: 27.42*

Today Dewey got started around 9:00AM. He walked on "Rails to Trails" for much of the time. However, he had to go thru a tunnel and slipped (icy). He had a hard time getting out of the situation. But God was his co-walker and led him through. How grateful we are for all the prayers and God's wonderful blessings. He will not walk on the trail tomorrow as there are several tunnels. He will walk on 50. He walked 27+ miles today. The trail was easy walking except for the tunnels. It was windy today but it was partly cloudy. Once again we've been blessed with good weather. We ate dinner at an average restaurant — service was poor — food okay. We talked with a couple at dinner and he was a retired truck driver. He told us there were 37 turns on Cheat Mt. I can believe it after driving it. Dewey will attest to it after walking it. We are grateful for God's watchcare today. Dewey's safety was His working.

Weather: partly cloudy and mild
Temp: 60's for a high

March 3, Wednesday-Day 23 *Miles walked: 29.53*

We got up early and grabbed breakfast at McDonalds. I took Dewey to the point for beginning his walk today. He did really well walking all day on hywy 50.-rolling hills but wide shoulders to walk on. He walked 29+ miles today and will get thru Parkersburg tomorrow and hopefully into Ohio. He has walked a total of 468+ miles. Pretty good for someone 27. I drove ahead to Parkersburg from Clarksburg (Bridgeport). I got us a room at Red Roof — very convenient as it is right on hywy 50. We will stay 2 nights and leave on Fri for N Ky and home for 2 nights. We ate dinner tonight at a Family restaurant — food good. I found a bakery in town that is like the old Latonia Bakery. Dewey likes donuts and they looked yummy. So I got him a half dozen. He and Jingle are enjoying them. It was nice and sunny this AM, but rainy in the PM. Rain is predicted for the next several days. We can't complain because God has been in charge of the weather and He has given us great weather. Thank you God.

Weather: partly cloudy and then rain
Temp: high's in 60's

March 4, Thursday-Day 24 *Miles walked: 29.76*

We got up early and ate breakfast at Mountaineer Family Restaurant. Dewey wanted to get an early start so he could make it into Ohio by late afternoon. He made great time today walking 29.76 miles. He will get closer to Athens, Ohio tomorrow and we'll leave for home in afternoon. I think 2 days off from walking will help Dewey's feet. They are sore and bloody at times. Jingle Bell has been a good little traveler. She hasn't walked much because of traffic on the roads; she is a "fraidy dog". Of course, I haven't walked any miles, but I've really driven lots! There's over 4,000 miles on the Tahoe now and we aren't back to KY yet. It was a dreary day today — some light rain — but basically pretty dry and cloudy. It felt good to cross the Ohio River today — leaving "Almost Heaven W VA" won't be hard to do. God continues to grant us blessings innumerable.

Weather: cloudy, light rain early
Temp: 50's

March 5, Friday-Day 25 *Miles walked: 16.24*
March 6, Saturday-Day 26 *Miles walked: 5.23*
March 7, Sunday-Day 27 *Miles walked:off day*

Dewey walked until 1:00 today and then we left Athens, Ohio for home. We will spend the weekend in Kentucky. On Saturday AM we attended the Ecumenical breakfast at Silver Grove Baptist. 18 of our folks were in attendance — wonderful support for Dewey and his walk. Dewey spoke re walk. Dewey walked 5 miles today and his total miles=518. In the PM, Ruth & Bill arrived to spend the weekend with us. They are always so supportive of us. We ate dinner at Karlo's. Ruth and Bill liked it. This morning (Sun) we went to church and had a wonderful crowd 50+ in attendance. Dewey preached on Suffering. He can relate to that in a physical way with all the pain of walking. Church hosted a delicious dinner after church. Good crowd stayed. We left around 2:00 for Athens taking the route Dewey will walk. We got to Athens around 4:30 to 4:45. Thank you God for safety and so many friends who care.

Weather: Friday-fair, Saturday-nice, Sunday-nice

March 8, Monday-Day 28 *Miles walked: 32.47*

We got up at 6:20AM and ate breakfast in the room. We will stay 2 nights at Days Inn in Athens, Ohio. Jingle is allowed to stay with us. Dewey began walking around 8:30AM in the snow flurries. He walked 32.47 miles today. He said it was a good day for walking. His ankle and leg are bothering him tonight. Probably overdid it today. He is so anxious to exceed his expectations. Tomorrow we will return to spot he stopped and pick-up once again. I shopped a little today and rested some in the motel. Dewey and I kept in touch through phones today. He had several calls today. These tend to keep his spirits soaring. We ate dinner tonight at Damon's. Food ok and service ok. Dottie South sent some pictures from Sunday. They were good — especially of the children. I feel so blessed. God has certainly answered the prayers for safety and no bad weather.

Weather: wet and snowy…no accumulation
Temp: 30's

March 9, Tuesday-Day 29 *Miles walked: 30.73*

Dewey got started early. We left Athens and went to spot where Dewey began walking toward Jackson. He walked 30.73 miles today. I couldn't find a motel to stay in and since we were within 80 miles of home we drove home and slept in our own bed. Jingle was happy to have her ramp again. Dewey has walked 580+ miles.

Weather: Sunny
Temp: low 40's

March 10, Wednesday-Day 30 *Miles walked: 32.39*

We got up very early to drive back to Ohio. Dewey began walking around 8:00AM. He began on Rt. 32 and turned onto Rt. 41 south. This ends at Aberdeen, Ohio where he will pick up 52 to the bridge over into Kentucky tomorrow. We ate lunch today in West Union, Ohio at Mikey's Family restaurant; it was good. He walked 32.39 miles today. He has now walked 612+ miles. We are staying at home tonight and will remain home until next Thursday.

Weather: fair
Temp: mid 40's

March 11, Thursday-Day 31 *Miles walked: 21.78*

We got up very early to return to Highway 41 in Ohio. Dewey walked 8+ miles down 41 to Highway 52 at Aberdeen on Ohio River. He then walked 2+ miles to the bridge which connects Ohio & Kentucky. We met Ray & Donna Throckmorton at Brown's Restaurant for lunch. Just as we finished, the Nelsons came. We left and went to the bridge and were joined by the press & radio station. Dewey was interviewed for radio and by newspaper. They took pictures and one reporter walked the bridge with Dewey, Ray and Bill Nelson. There were also 2 ladies present (from Ray's church) to cheer them on. Dewey got on Rt. 8 and walked 9+ miles. Ray met him and they returned to Maysville where Dewey stayed all night. Ray will take him back to Rt. 8 in the AM. I returned home for the night with Jingle. I will get a perm in AM and take Tahoe in to be serviced in PM. God has given us safety in abundance.

Weather: showers and then sunny
Temp: highs in 50's

March 12, Friday-Day 32 *Miles walked: 25*

Dewey started from Maysville this AM. Ray Throckmorton drove him to the spot at Dover and followed him all day. Sue, Charlie and Bill Nelson joined him for a 3+ mile walk. Also joining Dewey was Jim McCord. The Drewey's also came for a mile walk. I had blood work done, went to bank and grocery store and had a Perm in the AM. In the afternoon, I took the car to be serviced and then drove to Mentor to pick Dewey up. We came back by Camp Springs & ate at St. Joe's fish fry. Then we drove Dewey's route thru Campbell and into Kenton County. It is 26 miles from Melbourne to Walton UMC. Dewey walked 25 miles today. He's walked a total of 658+ miles. Tomorrow he will walk from Mentor to Melbourne and on to Camp Springs. The churches will be a part of the day and God will bless us as He has throughout the journey.

Weather: fair
Temp: low of 19, high of 41 degrees

March 13, Saturday-Day 33 *Miles walked: 17.22*

We got up early and met at the old Mentor Church (Baptist) in Mentor on highway 8. Several from the church joined Dewey to walk with him:

Garland, Willie, Patricia, Sarah, Kenny and Keith Daniels. Three others also came but did not walk. Dewey walked to Camp Springs and I picked him up and we went to Pendery Park to wait for those to come and walk to St. Philip's Church. A good group (about 36) walked with him including Darlene's dogs. Over 60 came to St. Philip's Church (Catholic) for the reception. Dewey was presented a plaque and March 13th was declared Dewey Sanders Day in Melbourne by the City of Melbourne. It was a nice affair. Maynard Futscher had made all the arrangements. He was as excited as could be about it. It was a beautiful day. God has smiled on us again. He is definitely in control. Dewey's sister, Martha, came and we were pleased about that. Dewey walked 17.22 miles today. His total is now 675+ miles.

Weather: fair
Temp: low 20's, high 40's

March 14, Sunday-Day 34 *Miles walked: 8.17*

We went to church at Melbourne United Methodist Church today. Ray Throckmorton preached. It was a meaningful worship time for us. After church Dewey walked 8.17 miles to the Kenton County line. He will walk up Visalia Hill and on to Walton in the AM. We were able to go to Brennan's 3rd birthday party-wild and wooly time with all the little ones. We relaxed at home after the party and watched NCAA pairings. Our UK Wildcats are the #1 pick. We then went to Cracker Barrel for supper.

P.S. It's been enjoyable to be at home for a few days. Nothing like one's own bed.

Weather: cloudy and light rain
Temp: 50's

March 15, Monday-Day 35 *Miles walked: 13.39*

It was a very busy day today. I drove Dewey to the starting point in Visalia at 9:00AM. Then I drove to Staffordsburg United Methodist Church to wait for him. They served coffee, donuts etc. Some ladies were quilting too. The DS and his secretary, Joyce, came to join us there. Dewey left walking with 4 from church with him. He walked past W.T. school and met the Cummins there to greet him. Also some came out from the school. Dewey went inside to greet some of the children. He then walked on to Craig's office. The Penns were there as was Charlotte Wethington,

Darlene and Sandy to walk with him. Craig had Dewey go inside and meet the staff. Dewey's nephew, Mark Fassler, also joined the walk. At the bridge in Walton, he was met by Jim Lawrence from Walton United Methodist Church. They walked to the church along with 2 ladies. The church served cookies and coffee and we had prayer with the group. I ran a shuttle service to and from the walk spot. Dewey walked on to the county line. He will begin there tomorrow.

Weather: beautiful day
Temp: 50's

March 16, Tuesday-Day 36 *Miles walked: 21.67*

Dewey began his walk in S. Walton around 8:15AM. He walked 21.67 miles today. He got to 2 miles south of Williamstown. Several walked with him today. David Bresser came and walked one+ miles; Kim and Craig walked 1+ miles also. Barry, Shelean, Tyler and Seth arrived around 5:30 PM. Kim fixed a good dinner for us. We also had birthday cake for Brennan's 3rd birthday. It snowed today but roads were clear. Dewey met a man who knew his dad on the railroad. He will go back tomorrow and take his picture as he forgot to do so today.

Weather: snowy
Temp: low 30's

March 17, Wednesday-Day 37 *Miles walked: 20.23*

HAPPY ST. PATRICK'S DAY

Shelean, Barry, Tyler and Seth walked with Dewey today. I ran the shuttle service to bring them back and forth. I made the round trip service three times (over 200 miles!). A dog they called Buddy walked with them for several miles. Dewey walked approximately 20 miles today. He walked to South of Sadieville about 1 or 2 miles. I took Greens and Darlene out to lunch at Skyline Chili. Then tonight we had carryout from the Court House in Independence. We ate at Kim and Craig's with extra folks joining us. Greens went to see Kim and Craig's new house. Hopefully, they will move in by the time we return. I'll miss my little guys. They will have changed so much by July! It's now 11:30 PM and we are still trying to get ready to leave in the AM. Thank you once again God for safety and manageable weather.

Weather: partly cloudy to cloudy, few snow flurries
Temp: 30's

March 18, Thursday-Day 38 *Miles walked: 18.56*

We got up early and finished the chore of packing the car. Shelean & Barry & kids left when we did around 8:00AM. I took Dewey to Sadieville and he walked until 11:00. We then went to Fava's in Georgetown to eat a bite of lunch. We then went to Lexington where Dewey was interviewed on WLEX (channel 18). This is his 2nd appearance on Ch 18. He returned to the walk and walked until 4:00. Dewey walked 18.56 miles today. We met Ruth & Bill for dinner at Applebee's in Frankfort and spent a little time at their house before going to Dewey's apartment in Versailles. We'll stay here until Monday AM. Ruth & Bill gave us a neat picture of Jingle Bell (a painting) as a gift to honor the walk. They are such special friends. They will join us in Owensboro for a couple of days and Ruth will walk some with Dewey. I got my hair done today and looked around the Outlet Mall in Georgetown. Dewey & I are tired tonight. Hope we sleep well. Thank you God for being in control and constantly reminding us of this fact. Jingle Bell will remain at home. We will miss her.

Weather: partly cloudy
Temp: 50's

March 19, Friday-Day 39 *Miles walked: 13.95*

BOOFIE'S BIRTHDAY

We got up at 6:00AM so Dewey could begin walking by 7:30. He walked until 8:30 and then I had to pick him up and come back to Versailles for a dental appointment to reseat a crown he lost early in the walk. Then he went back to Robertson Road and walked to McDonalds. Ann Crowe and Hester Souder walked from McDonald's to the Home (8 miles) with him. Dewey walked 13.95 (769+) miles today. We came back to his apartment to rest awhile. Ann & Hester organized the Tahoe and then we took them back to Mc Donald's to get Ann's car. We met Ruth & Bill at O'Charley's for dinner and went back to their house to watch UK's ballgame. Hooray! They won 95-75. The day was absolutely gorgeous; sunshine and rather mild. I can't thank God enough for all the blessings he's granted us on this trip. Dewey walked from Lexington to Versailles today.

Weather: sunny and mild
Temp: 50's

March 20, Saturday-Day 40 *Miles walked: 10.97*

Today was a special "homecoming" of sorts. A large crowd gathered at the Methodist Home to walk with Dewey. At 8:00AM he walked with Mayor Isaacs in Lexington. Then at 9:00 he walked through the gates at the Methodist Home where he was greeted by kids, staff and friends from near and far. They walked with him for under & over a mile. At 10:15 a National Guard helicopter landed as part of the celebration. The kids especially enjoyed this activity. Dewey, accompanied by an entourage, walked to the Court House in downtown Versailles. The mayor greeted him. He traveled by car to the Tyrone Bridge. The police escorted him across the bridge as it's very narrow with no walkway. Dewey walked in the afternoon almost to the bridge for a total of 10.97 miles today. We ate dinner with Ruth, Bill and Ann at Cheddars with a surprise visit from Shawn, Alysa and Eve…my first time to see Eve. She's a doll. Tomorrow we will visit Frankfort United Methodist Church.

Weather: cloudy with rain in PM
Temp: 60's

March 21, Sunday-Day 41 *Miles walked: 12.88*

We went to Sunday School with Ruth and Bill this AM at the Frankfort First United Methodist Church. Their Sunday school class has been very supportive of Dewey's walk. We shared some feelings and experiences of our venture. Dewey walked from bridge through Lawrenceburg-12.88 miles and then we met Julia and friends at Arby's for coffee. It was a nice reunion for us with Julia. She is very excited about Dewey's walk and plans to support it. We met Ruth, Bill, Ann and Hester at Bella Notte in Lexington for dinner. Great meal with special friends. UK lost their game so day ended on a sour note. Their next game will be in the fall. We spent the night in Versailles at Dewey's apartment and will travel to Bardstown for tomorrow night.

Weather: windy, cold and cloudy-some snow showers in PM
Temp: 20's in AM – 30's in PM

March 22, Monday-Day 12 *Miles walked: 32.16*

We got up very early and left for Lawrenceburg. Dewey had to meet a reporter at 8:00AM in Lawrenceburg. After the interview, he began walking towards Bardstown. He walked 32.16 miles today. Not much shoulder to walk on. Fortunately, traffic was pretty light until last 5 miles (closer to Bardstown). Carol Van Sickle called and she and Dan came over to Bardstown to join us for dinner at Dagwood's. It was a nice evening for us and a good perk for Dewey. There were no gas stations, stores etc. for much of the walk today. I took Dewey a light lunch around noon. He will meet with a TV reporter in the AM at 8:00. He wants to walk with Dewey and interview him. It's amazing how much we have been blessed during the walk. God has given us good health, safety and wonderful weather and also friends and family who care.

Weather: sunny & cold
Temp: 20's most of day, low 30's

March 23, Tuesday-Day 43 *Miles walked: 30.19*

Early to rise this morning. We ate breakfast at Stephen Foster Restaurant (buffet). Then at 8:00, Dewey met reporters and interviewed with them at the place he started walking today. Dewey walked 30+ miles today from Bardstown to Elizabethtown. He will walk with a minister friend for 11 miles tomorrow and his church is planning a light meal for us. They have collected $$ for the walk. I drove to find a place for us to stay the next two nights — Comfort Inn. The room has a Jacuzzi which Dewey needs and likes. I don't like tubs — showers are for me! We ate dinner tonight at a Steak House. Dewey got us ice cream for dessert. It was tasty. We'll go to bed early as Dewey needs to start early in the AM. Ruth & Bill will join us in Owensboro for the weekend. Our blessings are more numerous than we could hope for. Thanks again God.

Weather: sunny, not too cold
Temp: 30's for low, 50's for high

March 24, Wednesday-Day 44 *Miles walked: 26.04*

Another day has come and gone! We're spending our last night in Elizabethtown. Dewey walked 26+ miles today. The minister at Summit United Methodist Church and a parishioner walked also for 11 and 2 miles respectively. We had coffee with the folks at Summit in the PM. I did 2 loads of laundry today and got my hair done…about all I can say is it's clean. We ate dinner early tonight at a gift shop/coffee house called Back Home. It was quaint and the food was very tasty. Calls from Ken Sullivan and Sandy Clift. Sandy is doing a great job securing free lodging for us; 2 nights in Owensboro at Hampton Inn-great. We will stay 2 extra nights (Thurs-Sun). Ruth and Bill will join us in Owensboro on Saturday for the weekend. Thank you Lord for friends and for good health, good weather and safety. Answered prayer is always God's gift to us.

Weather: partly cloudy
Temp: 60's

March 25, Thursday-Day 45 *Miles walked: 32.21*

We left Elizabethtown early this AM after eating breakfast in the motel. Dewey started walking at 7:30(EST) and walked until 4:20 (EST) 3:20 (CT). He walked 32+ miles today on highway 62 and 54 and had very little shoulder so he dodged cars all day. He will walk thru Fordsville and on to Owensboro tomorrow. I came to Owensboro this AM and checked into Hampton Inn where we will stay for 4 nights. Sandy Clift got us two complimentary nights and we appreciate all her hard work and the generosity of Hampton Inn. We had dinner with Berry Major from Ky. Wesleyan College. He is in charge of Alumni Affairs at Wesleyan. We ate at The Bounty Restaurant. It is owned by former Ky. Wesleyan students. It was an enjoyable evening. We learned a lot about Wesleyan Alums and what's happening at Wesleyan today. Berry is very knowledgeable about the above. We had another lovely day for Dewey to walk. God is so good to us.

Weather: partly cloudy to sunny
Temp: high in low 70's

March 26, Friday-Day 46 *Miles walked: 28.04*

Today began early. Dewey walked 28.04 miles today. Judge Griffith who is on the Board at the home walked 2 miles with Dewey. An old friend of Dewey's from the past, Dr. Bill Tyler called Dewey. It was a nice perk for Dewey. He will try to come to church on Sunday to see and hear Dewey talk about the walk. Dewey found 2 pairs of shoes to use during the walk. He now has 7 pairs and, probably will need more. We ate dinner at the Moonlight Bar-B-Q. It was a feast. They have a marvelous buffet; even offer sugar-free desserts. Things will "pop" tomorrow at Mary Kendall Home. They plan a gathering and walk with Dewey. Ruth and Bill arrive tomorrow afternoon for the weekend. They will leave on Monday AM as will we. We go to Evansville on Monday; that's Indiana — another state on the journey. Our blessings are many and varied.

Weather: partly cloudy
Temp: 60's

March 27, Saturday-Day 47 *Miles walked: 8*

We got up early so Dewey could walk into Owensboro and be at Mary Kendall Campus by 8:00AM. The video crew wanted to interview Dewey re walk for a film to be shown at Conference this June. A good crowd gathered at Center of Hope to begin the walk into downtown Owensboro and over the bridge into Indiana. Friends from the past showed up to walk. Dewey's secretary from KWC, Brenda and her husband came. Also Chuck and Jane Ellen Epperson joined us after the walk. Balloons lined the route to the bridge so one couldn't miss where to go. Police escorted the entourage over the bridge (the 3rd such experience). We went to lunch with Chuck and Jane Ellen at the Cracker Barrel. We were their guests. They also gave us a sealed envelope (to use the contents for gas or food). Our friends are such treasures! Thank you God for showing us your love in this way. Ruth & Bill arrived and lunched with us. Chuck took us on a tour of KWC. Ruth, Bill, Dewey and I ate dinner at Moon Lite Barbeque. Such a feast!

Weather: partly cloudy
Temp: 60's & 70's

March 28, Sunday-Day 48 *Miles walked: off day*

Today we church hopped. Dewey had to speak three times today. Twice (9 & 11) at Wesleyan Heights and at Woodlawn United Methodist Church at 10:00. Hopefully, we will see monetary results from the effort. Ruth and Bill were with us all weekend. We lunched at Applebee's and suppered at Cracker Barrel. I'm eating too much. We toured the park area in Owensboro. Dewey, Bill and Ruth walked along the river walk. I read. The park area is all in bloom with pansies-quite colorful. We will leave in the AM for Evansville, IN. We saw many people today from the past, people I taught with at Washington Elementary and many Dewey knew from his time as a student at KWC.

Weather: sunny and mild
Temp: 60's

March 29, Monday-Day 49 *Miles walked: 28.82*

Today has been a day of "happenings"! Dewey started walking early in Indiana. I drove ahead to Evansville to get lodging for 2 nights. I got a nice room at Baymont Inn and was given a Jacuzzi room for price of regular room. I called Dewey to let him know and he related to me that Ray from Mary Kendall was walking with him for 2 miles and the radio and TV (channel 25) in area were interviewing him along the way. I stopped at a laundry which would wash and fold clothes. The man in charge, Tom Hall, knew all about the walk from TV coverage. He even knew Dewey Sanders. Laundry done free as contribution to walk. My sister, Shelby and people from their church came to Evansville today to cheer Dewey on. We lunched at Pizza Hut. Channel 7 (FOX) also came to interview Dewey. We had dinner at Angilo's in downtown Evansville with the Hoyes (Phil & Sandy). They treated us. They also knew the man who came over to talk to me in the donut shop. Small world — seems he is a kook but harmless. Sprinkled in AM then nice. God continues to control the weather.

Weather: partly cloudy and mild
Temp: 50's & 60's

March 30, Tuesday-Day 50 *Miles walked: 30.08*

Today we got up early and Dewey started walking a little after 7:00AM. I got a shampoo/set today while Dewey was walking. Shopped a bit in Dollar Tree and got the boys a few books & stickers. It rained today and Dewey had to change clothes at noon. He continued walking on 66 and it rained some more. By 3:00, he was drenched and chilled and called me to come get him. I of course did. He walked 30.08 miles today. Our friends from Henderson, Chuck and Jane Ellen Epperson, came over tonight and we shared dinner together at Logan's Roadhouse. The special thing happening for us is reconnecting with friends from long ago. They have gotten interested in the walk and plan to talk it up at Henderson United Methodist Church. Isn't it amazing how God works! Many have recognized Dewey from seeing him on TV. They've honked and one man stopped and gave him $5.00 for the cause. Oh — the wonder of it all!

Weather: rainy and cool
Temp: 50's

March 31, Wednesday-Day 51 *Miles walked: 28.83*

We got an early start this morning after packing the car and eating a bite of breakfast. Dewey began walking around 7:30 and walked to New Harmony, IN. It's an interesting town right on Wabash River. I drove ahead to check out the bridge and found it was non-pedestrian. I went back to a Food Mart and asked if I could call someone re an escort over the bridge. She called and got permission for Dewey to walk over the bridge and they didn't even charge us the $1.00 toll. Two different folks gave Dewey money for the walk ($30.00 total). I got a room at Best Western in Grayville, Illinois and when they found out about the walk they gave us free lodging and free dinner and breakfast. We are truly the most blessed folks in the world. Someone came to interview Dewey for the local paper. It just doesn't stop! God continues to give us blessing after blessing. It rained today but Dewey was able to walk 28+ miles today. He is now over 1,000 miles as of 10:00AM.

Weather: cloudy and some showers, peeks of sun at times
Temp: 50's

April 1, Thursday-Day 52 *Miles walked: 25.33*

Dewey hit the road early right after we ate the free breakfast in the motel restaurant here in Grayville, Ill. Dewey walked 25.33 miles today. He stopped a little early because he was chafed. I kept up with him somewhat today because there were no services on Rt. 130 (except for Albion) all the way to Olney. Dewey will start at that point tomorrow. We ate dinner in the hotel again. Food pretty tasty. Dewey will work on computer tonight. I'll watch some TV. It's been a calm day today. We'll stay at Effingham, IL. tomorrow night.

Weather: partly cloudy and cool
Temp: 50's

April 2, Friday-Day 53 *Miles walked: 28.19*

We left Grayville early and I drove Dewey to the beginning point for today. He walked 28+ miles today going almost to Wheeler. I drove ahead to Effingham where we will stay for 2 nights. I got us a room at Comfort Suites and it's nice. They serve continental breakfast. We ate dinner at Nienerg's Family restaurant, food good-great desserts-even sugar free. We wrote thank you notes (cards) to contributors-still have many to go. The Motel (Best Western) in Grayville gave us 2 nights lodging free as well as a breakfast & dinner. We could not believe their generosity. God continues to bless us each and everyday in many ways.

Weather: nice
Temp: 50's

April 3, Saturday-Day 54 *Miles walked: 32.16*

We got up early and ate breakfast in the motel. I drove Dewey back to several miles past Wheeler. He walked a total of 32+ miles today. His total now is 1121 miles. Quite a feat I'd say. I did laundry and wrote some cards. Dewey took a break for lunch and I got him a grilled cheese and homemade banana cream pie. I ate a salad & SF cherry pie. This afternoon Dewey walked again and I watched a little TV and took a nap. We ate dinner at the Niemerg's Family restaurant again tonight. Ate in dining room this time and had salad/soup bar and dinner items. We will leave in AM for Dewey to speak in Hillsboro, IL. We plan to stay in Litchfield, IL for 3 or 4 nights. I will have to run shuttle service the next few days. Weather nice today, but wind was high. Thank you God for watching over us again today.

Weather: sunny, cool and windy
Temp: 50's

April 4, Sunday-Day 55 *Miles walked: Off day*

PALM SUNDAY

We got up early, ate breakfast and packed the car. Dewey won't walk today but we'll drive the rt. to Hillsboro, Illinois which he will walk on Monday & Tuesday (approx 50+ miles). We went to church at 10:00AM at Hillsboro United Methodist Church. The choir entered with palm branches. Service was nice. We left after church and drove to Litchfield (8 miles w) to secure lodging for next 3 nights. We're staying at Comfort Inn. We have a nice suite w/Jacuzzi for Dewey. At 1:00PM we returned to Hillsboro United Methodist Church for lunch and meeting for Dewey to talk about the walk, about 20 or 25 in attendance. Dewey's friend from college, Charlie Sensel, and his wife drove in from Rockford, Illinois to be a part of the day. Two churches got together to sponsor this-food and fellowship great. It was a "beautiful day in the neighborhood". Mr. Rogers would have liked it. Blessings continue.

Weather: sunny & cool
Temp: 50's

April 5, Monday-Day 56 *Miles walked: 30.11*

The day began early as we had to travel almost 40 miles back for Dewey to begin walking. He walked 30.11 miles today on fairly desolate country roads, little towns like Ramsey, Fillmore etc. I took him a light lunch midday. We ate dinner tonight at Ariston Café, a family owned restaurant since 1924. It is on Route 66 and there is the old historic Rt. 66 nearby. Dewey will walk that Rt. for 8 miles tomorrow to Rt. 108. He'll get some kicks out of walking on old route 66. The weather has been glorious, bright sunshine and mild temperatures. How blessed we are! God is our constant in this whole journey…good health, good weather and safety…answered prayer is evident day after day.

Weather: gloriously mild
Temp: 50's-60's

April 6, Tuesday-Day 57 *Miles walked 25.12*

We got up early as usual and had breakfast and I drove Dewey to the place for him to start walking. He walked into Hillsboro and met me at the laundromat where I stopped to do our clothes. Dewey had a dental appointment at 11:30 to have a crown repaired. We then went to Hillsboro United Methodist Church to meet with some folks to walk thru town with him. Two newspapers came to take pictures of the group. I met them at Sevdali's restaurant to eat lunch. Dewey continued walking after lunch to Litchfield. He walked 25.12 miles today. He will walk on Rt. 66 tomorrow. We ate dinner again at Ariston Café. Food is great there and atmosphere is nostalgic. Dewey e-mailed the newsletter to Ann tonight for editing. She will forward it on to Bill to be sent out. We are so grateful for all the blessings and joys we have received everyday. God is ever present with us.

Weather: warm and sunny
Temp: 60's & low 70's

April 7, Wednesday-Day 58 *Miles walked: 32.03*

The day began earlier for Dewey than me. He started walking on Old Rt. 66 right by the motel (Comfort Inn). He walked thru Carlingburg and on near Carrolton. I had a call on my cell phone from Carol Culbertson. She and Steve would arrive to surprise Dewey (Steve is Dewey's cousin). He will walk a few miles with Dewey tomorrow. We will leave after they walk and go on to Hanibal, MO for the next 3 or 4 nights. Dewey will back track. There are no other motels closer. Dewey will walk Thursday and Friday and take Saturday and Sunday off. He needs a break to rest his feet, legs and back. We ate dinner at Ariston Café again tonight. Steve and Carol enjoyed it too. I drove many miles today checking out terrain and services. No service and no shoulder on the roads. Dewey will need to be careful. Weather is still unbelievably lovely, sunny to partly cloudy and mild temperatures. Again we say thank you God for blessings immeasurable.

Weather: mild & sunny
Temp: 60's & 70's

April 8, Thursday-Day 59 *Miles walked: 31.55*

Our visit with Steve and Carol (from Wis) was a pleasant one. They both have a wonderful gift of gab. Carol is a storyteller re pioneer women (for schools & clubs). Steve walked 8 miles with Dewey and then Dewey walked on and Steve & Carol left for home. Dewey walked 31.55 miles today. We had to take a ferry across the IL River…no other way to cross at this point (many miles out of way to walk). Dewey will begin walking tomorrow at Mozier and walk up IL 96. We are staying at Comfort Inn in Hanibal, MO for the next 5 nights. Dewey will back track as there are no other places to stay along the way. I was pleased to receive a call from Dottie South this morning. It is always a perk to hear from some one back home. I keep exclaiming about the weather, but it has been so lovely for most of the trip. God's wonderful gift to us…answered prayer from our many prayer warriors coast to coast. We ate dinner tonight at Mark Twain Diner. The last time we ate there, Mother was with us (2001).

Weather: pleasant but windy
Temp: 60's

April 9, Friday-Day 60 *Miles walked: 32.42*

GOOD FRIDAY

The day began early as Dewey had to return to starting point some 50+ miles away. He walked 32.42 miles and will begin at Atlas, IL on Monday. He will not walk for 2 days except to cross the Miss. River early Sunday morning. I stayed fairly close to him today as his back was bothering him. If he needed anything, I'd be close by. I read quite a bit today and we lunched at a coffee shop near Atlas where walk will begin on Monday. There was a Country Store there and I looked around but didn't buy anything. We ate dinner at Logue's Restaurant which is near the Comfort Inn where we are staying. Weather was good again today, but windy. Our blessings are so many it's difficult to take it all in. God is truly Good all the time.

Weather: sunny and windy
Temp: 50's & 60's

April 10, Saturday-Day 61 *Miles walked: Off day*

Dewey took the day off from walking to let his back rest. He will also take tomorrow off. We plan to go to a sunrise service at 7:00 and to a brunch at the Lula Belle's Restaurant in downtown Hannibal, MO. Dewey rested a lot today. I took laundry to get done. Then I shopped at Walgreen's for a couple of items Dewey needed.

Weather: rainy and cool
Temp: 50's

April 11, Sunday-Day 62 *Miles walked: 1.4*

HAPPY EASTER

We got up at 5:45AM for Dewey to run the bridge. Everything worked out ok. We then attended a sunrise service at Immanuel Baptist Church. It was a good service and then we had breakfast with them. We had lunch at Lula Belle's Restaurant and B&B. It was good too. Tonight we ate at the Pirate's Cove restaurant. Guess all we did today was eat! There was a gorgeous sunrise this AM. God's signature for Easter — Alleluia.

Weather: sunny and cool
Temp: 50's

April 12, Monday-Day 63 *Miles walked: 32.15*

Dewey finished walking in IL today. He walked 32.15 miles and his total is now 1200+++miles. I took the Tahoe in to be serviced today. It's good for another 3,000+ miles. I checked on Dewey some today and watched a little TV. We ate dinner at Fiddlesticks. It's a lot like Friday's or Applebee's.

Food and service good — atmosphere quaint. We'll be leaving tomorrow for Macon, MO where we'll stay for 2 days. It was a breezy day today. Dewey really made good time as the wind was behind him for most of the day. Dewey will be walking on US 36 for many miles beginning tomorrow…across MO, Kansas and part of CO. Thank you God for blessing us yet again.

Weather: sunny and cold
Temp: 30's to low 50's

April 13, Tuesday-Day 64 *Miles walked: 32.09*

Dewey began walking early around 7:30AM. I drove on to Macon, MO to find lodging for the next 2 nights. I found a new Comfort Inn and got a lovely room w/Jacuzzi for price of a standard room. People have been so nice to afford us this "perk." I checked out restaurants and we ate at one called "The Maples" which was a really nice place to eat; our kind of eatery. Dewey walked 32.09 miles today and is now well over 1300 miles. A lady from the local paper came and interviewed Dewey for the paper. I think the paper's name is Journal. We'll be here until Thursday. Can't wait to taste the breakfast in the AM…understand the waffles are really tasty! Beautiful day today, but windy. It was a hard day of walking because of the wind. Blessings continue.

Weather: sunny and cool
Temp: 29 degrees early up to 50's

April 14, Wednesday-Day 65 *Miles walked: 32.24*

Comfort Inn here in Macon, MO served a great breakfast, waffles etc. We enjoyed it. Dewey talked with Jack Pattie (Lex. D.J.) this AM early. He walked 32.24 miles today. His total is 1390+. I shopped some today but didn't buy much; 2 cards and some envelopes. The weather continues to be spectacular. The Red Sea opens and gives us dry land each day. What more could we ask for? God answers prayer. We are grateful for the blessings each day affords. We have enjoyed the amenities of this motel. It has only been open since 1st week in Feb. Everything is fresh and new. I picked Dewey up this PM and he enjoyed the Jacuzzi before we went to dinner. We ate at a family restaurant tonight. Food was okay.

Weather: sunny and nice
Temp: low in 30's, high 70's

April 15, Thursday-Day 66 *Miles walked: 30.17*

TAX DAY

We had to leave the lovely haven in Macon, MO. We enjoyed our two days/nights there. I drove ahead to Chillicothe, MO. to find lodging for tonight. Not much to choose from; staying in a Motel 8. It is ok but not like we're used to. I also had a hard time finding a decent place to eat. We're going to bed early. Dewey needs a good night's sleep. I do too. God is always watching over us.

Weather: sunny and mild but windy
Temp: low in 30's, high in 70's

April 16, Friday-Day 67 *Miles walked: 32.14*

We arose early for Dewey to get started and get closer to Cameron, MO. I drove ahead to secure lodging for tonight. We already had reservations for Saturday and Sunday (thru Sandy Clift) in Cameron. The Holiday Inn Express is a nice motel with deluxe breakfast. When we arrived at the motel there was a big basket of fruit, 2 balloons, $40.00 in food coupons at Ma & Pa's Kettle and 3 angel bears (wings & all) and I just thought of the Trinity; Father, Son and Holy Spirit. This was a gift of love from our Church, Melbourne United Methodist Church. They just keep on giving. Thank you God for such caring folks who love us and You very much. Another beautiful day. Dewey walked 32.14 miles today. His total is now over 1450+ miles. It was actually hot today. It reached over 80 degrees.

Weather: sunny and warm
Temp: high in 80's

April 17, Saturday-Day 68 *Miles walked: 28.82*

Dewey began walking early to "beat the heat" so to speak. It got up in the 80's today. We ate lunch at Ma & Pa's Kettle again and Dewey continued walking on toward St. Joseph, MO. We drove into St. Joseph today to check out the bridge over the MO River into Kansas. Dewey will walk that tomorrow in the afternoon. Dewey talked to the Webbs & Roots in Corbin today. Larry is tracking Dewey on U-Locate. Blessings continue to mount.

Weather: warm and windy
Temp: 80's

April 18, Sunday-Day 69 *Miles walked: 1.4*

We got up early and went to Cameron United Methodist Church at 8:30AM. The folks at the church were really great. The service was meaningful. At the end of the service the minister presented us with the "mission basket" for the walk. I counted the $ this afternoon and the total was $88.53. We then went to eat at Ma & Pa's Kettle. One of the church members (a state lobbyist) bought our lunch. We just keep on being blessed. We drove back over to St. Joseph and Dewey walked the bridge. It was terribly windy! Dewey's total mileage is now 1479+ miles. We move on tomorrow and will stay in Kansas. Guess I'll play Dorothy from Wizard of Oz and click my heels for a magical drive across Kansas!

Weather: very windy, cloudy, few rain drops
Temp: 80's

April 19, Monday-Day 70 *Miles walked: 29.89*

We left Cameron, MO this AM. The motel (Holiday Inn Express) gave us the greatest deal! One free night and the other two nights at $29.00 per night. Blessings continue. We are staying in St. Joseph, MO tonight at a Drury Inn. It looks like they're going to serve a nice hot breakfast. Dewey walked 29.89 miles today and is 10 miles into Kansas. We will stay at Hiawatha for the next two nights. Tomorrow David Bresser is going to walk some with Dewey. We'll probably meet at Rt. 7 in Troy, KS in the AM. We ate lunch today in a Bar-B-Que restaurant called Bandanas. It was tasty. Just ate Wendy's for dinner. Weather great today but windy and not as warm as it has been. God is a very present help to us. Each day brings new blessings galore.

Weather: sunny, windy and mild
Temp: 60's & 70's

April 20, Tuesday-Day 71 *Miles walked: 30.76*

Today was a special day for Dewey. David Bresser came up from Ft. Leavenworth, Kansas to walk with him. They walked 12 miles together. David is a Captain in the Army and was our neighbor when we lived on Campus Drive. We all ate lunch together in Troy at Carolyn's Restaurant. Food was tasty. Dewey walked 30+ miles today and is now at 1538 miles. I trudged ahead to Hiawatha, Kansas to find lodging for the next 2 nights.

We're staying at the Hiawatha Inn. It was raining early this morning, but by the time we got to Dewey's starting point for the day it was just a drizzle and then it stopped and sun came out until late in afternoon when it rained again. God opened the Red Sea once again and Dewey walked on dry land! Isn't it wonderful how God provides? Words are inadequate to express my gratitude.

Weather: rainy, cloudy, sunny, then rainy
Temp: 60's

April 21, Wednesday-Day 72 *Miles walked: 32.15*

Early to bed, early to rise once again. Dewey began close to the motel and so I was able to shower and wash my hair before taking off today. I checked on Dewey via phone and joined him at the Cozy Café in Fairview, Kansas. He had met Mel and he insisted we come into the Café and meet the people. Pat Weaver came from the local paper and talked to us re walk. We plan to stop at the Cozy Café for breakfast in the AM. Dewey walked 32.15 miles today under sunny skies and in mild temps. We ate dinner at the Restaurant here in the motel. The food is good, but atmosphere is plain and simple. Blessings continue to come our way. Thank you God for your watchcare of us all along the way. Kansas has proved to be more hilly then expected. Another sign of variety in God's universe.

Weather: sunny and mild
Temps: 60's, low 70's

April 22, Thursday-Day 73 *Miles walked: 32.26*

We got up at 6:00 AM so we could pack the car, check-out and leave for Fairview to eat breakfast at the Cozy Café. We enjoyed visiting with the locals and we received a free breakfast. What nice folks we've met in Kansas. Dewey walked 32.26 miles today and we're staying the night in Marysville, at Oak Tree Inn. There's a Penny's Diner Restaurant right in front of motel and we ate lunch there today. This evening we enjoyed a nice dinner at the Koester House which is in an old antebellum type home that goes back to the beginning of 20th century. I had grilled shrimp and it was delicious. We will move on 50 miles tomorrow to Belleville, Kansas and stay there 2 nights. It means I go back and forth driving but allows us to stay 2 nights in same place. It rained today but Dewey walked anyway…no thunder or lightening just steady rain. God knows the farmers need rain. Thank you God on their behalf.

Weather: rainy & cool
Temp: 50's

April 23, Friday-Day 74 *Miles walked: 31.81*

Our day began at 6:30AM. We packed the car, ate breakfast and were on our way around 7:30. Dewey walked 31.81 miles today along 36 thru Washington, Kansas and got within about 10 – 15 miles of Belleville, Kansas. He will get thru Belleville tomorrow and on past 15 miles. We will stay two nights here at a Best Western. We ate dinner tonight on the square in Belleville at AJ's café. Food was home cooked and the servings were too generous. Neither one of us could eat it all. Dewey has been unable to get on the internet and it's frustrating for him. We finally figured out how to retrieve the voice mails on the cell phones. Dewey had piled up 16 voice mails. Weather was cloudy today but no rain. Dewey felt it was a good day for walking. Our blessings continue to mount! How grateful we are.

Weather: cloudy and cool
Temp: 50's

April 24, Saturday-Day 75 *Miles walked: 15.19*

The heavens opened this morning and it was cold so Dewey waited until afternoon to walk. He was able to walk 15.19 miles today. His total is 1648+ miles. He plans to walk again tomorrow to make up for loss of walking time today. We will go to church and then he will walk. We over ate today. AJ's Café is too tempting. We ate there 3 times today! Dewey can't get his computer to work so he went to library and got on line to get e-mails. I read while he did that. We're staying in Bel Villa Best Western in Belleville, Kansas. It is nice and convenient; right on 36. Today was rainy and dreary. But God knows rain is needed in the area. No severe thunderstorms, just wet. We have so much to be thankful for. God is ever present with us on this journey of a life time.

Weather: rainy
Temp: 40's – 50's

April 25, Sunday-Day 76 *Miles walked: 19.92*

We enjoyed another glorious Sunday with Methodist folk! We visited 1st Methodist Church in Belleville, Kansas. Dewey was recognized and his walk emphasized. It was confirmation Sunday and the bell choir did special music. There was a dinner after church and we were invited to join in the fellowship. It was nice and we felt like we were home at Melbourne United Methodist Church. Some of the church folks contributed to the walk over $100.00. Dewey had several calls today from unexpected sources. The calls were from Joe Roop, Chuck Epperson, Jim McCord also from Janet & Lowell May from First Methodist in Belleville, Kansas. Dewey walked 19.92 miles today. It was a beautiful day in Central Kansas. Tonight we ate at AJ's Café. We've been regulars there for the three days we have been here in Belleville. Thanks be to God for granting us another special Sunday. Methodist fellowship is certainly universal.

Weather: sunny & cool
Temp: 50's

April 26, Monday-Day 77 *Miles walked: 32.24*

HAPPY BIRTHDAY, MAMA DOR!!

We ate breakfast at 7:00AM with Janice & Lowell May at AJ's Café. Janice is from Kentucky. We then returned to the motel to pack and get "on the road". Dewey walked 32.24 miles today in ferocious wind (30 to 40 miles per hour). We saw the point which is the geographic center of the contiguous (48) states. We are now at the halfway mark of the walk. We are staying tonight at the US Center Motel in Smith Center, Kansas. We ate dinner at Dave's Café next to the motel and the food was okay. We'll probably grab a bite of breakfast there in the AM before Dewey returns to point of walk. We will stay the next two nights in Phillipsburg and we hope to exit Kansas by end of week or first of next week. The weather was sunny and cool but the difference was the high wind. Dewey was really tired tonight after fighting the wind all day. We continue to experience the gift of good weather. God is truly a very present help.

Weather: sunny, cool and windy
Temp: 50's – 70's

April 27, Tuesday-Day 78 *Miles walked: 32.33*

Our day began at 6:30. We got dressed and went to breakfast at Paul's Café. Then we packed the car and I took Dewey to the starting point for the day. I drove to Phillipsburg and secured a room for next 2 nights at Cottonwoods Motel. I went into downtown Phillipsburg to look around. I went into a card shop and the ladies working there were interested in Dewey's walk and called the local paper. Reporters came and talked to me and then went out on highway 36 and interviewed Dewey and took pictures. The day was beautiful; sunny and mild. God answered prayer once again giving us dry land for walking and driving. Dewey walked 32.33 miles today for a grand total now of 1731+ miles. We are getting there, over half way under our belt. Praise, Praise, Praise. We ate dinner in a restaurant next to motel; food okay.

Weather: sunny and mild
Temp: high 70's

April 28, Wednesday-Day 79 *Miles walked: 32.61*

Dewey began walking at 7:00AM today. He wanted to get in as many miles as he could before it got too hot. (It reached 90 degrees today.) We ate a late breakfast "on the road" and I went and got a haircut. My hair feels better and looks better too. I've been having many bad hair days! Dewey was able to quit a little earlier today. He walked 32.61 miles and was tired from the heat by the time he stopped. I did laundry while Dewey went to library to check e-mails. He is unable to get on line right now; no service here in mid-Kansas. We ate dinner in the Branding Iron II tonight as we did last night. I was touched today when the girl donated my $18.00 hair cut to the walk. It was unexpected and I put the $18.00 in the donation envelope. We have had so many "out of the blue" happenings. I received a call from Beulah Lykins today and it was a special treat to talk to someone from home. What a beautiful day we had today. It got hot but the big sky was gorgeous. God has blessed us yet again; thank you, thank you, thank you.

Weather: lovely, cool AM, hot PM
Temp: It got to 90 degrees today

April 29, Thursday-Day 80 *Miles walked: 30.7*

Today was a cold, windy, rainy (off & on) day. Dewey trudged through the elements off and on as well for 30.7 miles. I drove & drove trying to find a decent place to stay on 36 but to no avail. Finally, had to drive south 30 miles to Goodland, Kansas where there were nice places to stay off I-70. This means back tracking but it's better than staying in "spacious" again. The motel we are staying in is a Best Western and there is a restaurant attached to it. We drove into town after we ate dinner looking for other places to eat. We found very few. We did find Methodist Church which we will attend on Saturday at 5:30 or Sunday AM. We entered Mountain Time zone and are now 2 hours behind Eastern Time or Kentucky time. We hope for better weather tomorrow and higher temps. It has been cold for us, 30 & 40 degrees for highs! We still thank God daily for His ever presence with us.

Weather: cold, rainy (off & on) windy
Temp: 30's & 40's

April 30, Friday-Day 81 *Miles walked: 31.66*

It was back to winter clothes today! Temperatures were in the 30's most of the day. There was ice on the car hood this AM. We are staying in Goodland, Kansas until Sunday. There aren't very many places to eat in Goodland. We ate at Hank's which is a Mom/Pop type restaurant. It is really hard to find vegetables. I've had to stick with salad bars for the most part. Dewey has walked 1824+ miles so far. We hope to be in Colorado by Monday. I've enjoyed Kansas but it's been a long journey across the state. "Rocky Mountain High" is on the horizon. I took a long nap today and just hope I'll be able to sleep tonight. Guess I'm more tired than I've realized. Since we are now on Mountain Time we have another adjustment to make; maybe that is why I'm tired. We are grateful for all our blessings.

Weather: cold and rainy
Temp: 30's

May 1, Saturday-Day 82 *Miles walked: 31.92*

MAY DAY & DERBY DAY

Dewey started walking early. It was a sunny, cool day, great for walking. Dewey has now walked 1855+ miles. We will leave here tomorrow and go on to Yuma, Colorado to stay for 2 or 3 days. I will do back tracking but there just aren't many places on 36 or 27N into NEB and on to 34 in Colorado. We watched the Derby and were glad that the favorite won for a change; yeah, Smarty Jones. We went to church tonight at 5:30 at First Methodist in Goodland. Their minister is a woman, Janice Hernandez. It was our first time to take Communion "on the road". It was meaningful to "break bread" with Methodists in a different place. After church, we went to Wendy's with minister and a couple from the church. We enjoyed the fellowship. God has given us many meaningful times of worship — to

God be the glory, great things He has done.

Weather: sunny, cool then mild
Temp: high in 40 to 50 degrees

May 2, Sunday-Day 83 *Miles walked 32.04*

Since we went to church last night, we got up early, ate breakfast, packed the car and got on the road so Dewey could get an early start walking. He walked 32.04 miles today and his total is now 1887+ miles. Tomorrow he will walk the last few miles in Kansas and 8.5 miles in Nebraska. He will cross over the state line into Colorado. We are staying 2 or 3 nights in Yuma, Colorado at the Nelson Inn (not Willie)! It is a new motel and nice, but there aren't very many places to eat in Yuma. We ate dinner tonight at Woody's (a hamburger place). We will probably have to eat there another time or two. Weather was lovely today…sunny and mild. Guess we will be early to bed, early to rise again tonight. We are so blessed; good health, safety and good weather continue to be our gifts from above.

Weather: sunny, cool to warm
Temp: 70's

May 3, Monday-Day 84 *Miles walked: 31.79*

We are in Colorado both in sleeping and Dewey's walking! John Denver's "Rocky Mountain High" is upon us as we feel the altitude rising with each step Dewey takes and each mile I drive. We will be one mile high before too long. Dewey walked from early morning in Kansas, Nebraska (for 8.5 miles) and finally in Colorado. He stopped for a lunch break and we ate at a quaint café (Canyon Coffee and Café). We really liked it; our kind of place. Dewey walked again after lunch for a total today of 31.79 miles for an overall total of over 1900 miles. We are getting there! We did laundry late this afternoon and grabbed Pizza Hut for dinner. Not many eateries in this town of Yuma, Colorado. The motel is very nice, brand new in February. God continues to meet our needs and we "praise God from whom all blessings flow." AMEN

Weather: sunny and nice, hot in PM
Temp: 80's

May 4, Tuesday-Day 85 *Miles walked: 32.33*

The day began early for us again this AM. Dewey was out walking a little after 7:00AM. It is cooler in the AM. Temps in 80's by afternoon. I picked Dewey up around noon for a break. We ate lunch at the Country Club restaurant, Sweet Spot Café. Food was good. It isn't open tonight. Dewey continued walking after lunch. He walked 32.33 miles today. Altitude is gradually going up. Dewey is noticing it as he walks. I don't notice it yet. The Feeding Yards are a sight to behold! The smell is like nothing I have experienced before. I don't think I'll ever want to eat the corn fed beef again. It almost persuades me to become a vegetarian. We ate dinner at Pizza Hut again tonight, even had fresh strawberries to put on ice cream. Dewey has now walked 1950+ miles. We move on tomorrow to Ft. Morgan, Colorado. God remains the constant in all of this.

Weather: sunny & mild
Temp: high in 80's

May 5, Wednesday-Day 86 *Miles walked: 30.69*

We have become early risers! It was 6:00AM again today. We had to pack the car and leave early so Dewey could walk while it was still fairly cool. Temps got in 90's out here in the day and the heat is draining but Dewey still walked 30.69 miles. I ran a shuttle service today because I had to drive 50+ miles to get lodging for the next 2 nights. I got a room at Days Inn in Fort Morgan, Colorado. Dewey was interviewed twice today by newspapers, one in Akron and the other in Ft. Morgan. He is becoming quite the celebrity! His head is getting so big the Crocodile Dundee hat will hardly fit; ha! Dewey has walked a total of 1980+ miles. We ate dinner tonight at a restaurant called Memories. Food was tasty! At least it wasn't meat and potatoes. Dewey has to return east of Brush, Colorado to walk tomorrow. He hopes to walk thru Ft. Morgan and beyond. Sherry Fuller called today. It was nice to hear from home. Blessings continue to mount. Thank you God.

Weather: sunny and hot
Temp: high of 90

May 6, Thursday-Day 87 *Miles walked: 32.52*

Our day began really early for us this AM. Dewey was walking by 7:00AM. We left at 6:30 to take Dewey to his starting spot just east of Brush, Colorado. He walked 32.52 miles today and will begin walking tomorrow on 34 at Wiggins, Colorado (which is west of Ft. Morgan toward

Greely). It was very hot today. By afternoon the temperature was in the 90's. We were glad it was cooler earlier this morning. Dewey made good time early on. He has walked a total now of 2012 miles. We have passed the 2,000 mark. Newspaper photographer met Dewey early to take pictures. Maynard Futscher, Father Cooney and a group from St. Philip Church in Melbourne called tonight. What a surprise! Isn't it wonderful how hearts reach across denominational lines...another pick-us-up. We ate dinner tonight in an Italian Restaurant called Cables Italian Grill. It was good. Praise God from whom all blessings flow!

Weather: sunny and hot
Temp: high in 90's

May 7, Friday-Day 88 *Miles walked: 32.25*

5:20 came early. Dewey called Jack Pattie in Lexington (Radio DJ) and because of time difference (2 hours) the call had to be at 5:40 MT. We did get away early even with packing the car after 2 days in Ft. Morgan. Dewey started walking at 7:00am at Wiggins. He walked to Greely (32.25 miles). I was unable to get lodging in Greely due to a college graduation on Saturday; therefore, I drove to Loveland, Colorado and got lodging at a Hampton Inn. There was a Black-Eyed Pea Restaurant near the motel. It's a favorite place of ours to eat. We enjoyed it immensely. Dewey has walked a total of 2044+ miles, quite an accomplishment; only 1500 more miles to go. The Rockies and the desert are ahead of us! We will move on to Ft. Collins tomorrow even though Dewey won't be there yet in the walking. God is still very real to us. We are so blessed every day. The Mitchell sisters called Dewey today to encourage him. They come to our homecomings at Melbourne.

Weather: sunny and hot
Temp: reached 90 degrees again

May 8, Saturday-Day 89 *Miles walked: 32.1*

Whew! Day began early and Dewey started walking a little after 7:00. He walked 32.1 miles today to within several miles of Ft. Collins, Colorado. I had my hair done this AM and it looks okay, not Diane okay, but okay. At least it is clean and she used a curling iron so maybe I'll be able to keep it for a few days. I got us a nice place to stay in Ft. Collins. It is a Quality Inn and Suites. We will probably be here until Thursday AM. Motels were booked up for tonight due to a ball tournament. God continues to meet our needs. Thanks, thanks, thanks. Dewey has walked a total of 2076+ miles. That is some feat. We ate dinner tonight at Nate's which is a Seafood/steak restaurant. Food was excellent and service too. Jim Smith will join us on Tuesday to drive Dewey some. This will be a nice break for me. Temperatures continue to be above average for this area but should moderate by Monday.

Weather: sunny and warm
Temp: high in 80's

May 9, Sunday-Day 90 *Miles walked: Off day*

HAPPY BIRTHDAY & MOTHER'S DAY TO ME

What a day it has been! We got up early and ate at Egg and I, then went to church at Christ United Methodist Church at 8:00 am, traditional service and formal. We had coffee after church with church folk. We came back to motel and Dewey cleaned out the car (also washed it on outside). The church (Melbourne United Methodist Church) called to wish me a happy birthday (in song). Then the Pryors (my friend Phyllis Reed) from Longmont, Colorado came up and took us to lunch at County Buffet. Enjoyed reconnecting after 38 years! Longtime-no see. They left about 3:30 and Dewey and I rested. Others called with birthday wishes; Dorothy Cummins, Peggy Leonard, Ruth Derrington and Hester Souder. Dewey got me cards for birthday and Mother's Day and 3 pieces of sugar-free candy. I'll get a gift later; that's the way I wanted it. The kids also called with wishes. I was on the phone a lot. We ate dinner at Bennigan's; too much eating for one day. I'm grateful to God for all His blessings to me everyday.

Weather: Sunny and mild
Temp: Low 80's

May 10, Monday-Day 91 *Miles walked: 28.17*

A DAY TO REMEMBER

God is the giver of miracles. He gave us some today. Dewey changed his walk route by moving ahead since Jim Smith is arriving tomorrow and we want to be closer to motel. In so doing, we met a team of marathon runners from Boston, MA running for children's charities. They were interested in Dewey's walk and one of the guys is the head of the Boston

Marathon. He told Dewey to consider himself entered for next year's Boston Marathon. Dewey is elated! Dewey walked 28.7 miles today. Then I took the car to be serviced and a man walked in with me. He asked about the sign re walk. I explained what Dewey was doing and he was impressed! When I took the car back at 4:00 for actual service, I was told the service was free. The man I walked in with in the AM was the owner and he said NO CHARGE! God truly works in mysterious and providential ways! The Rockies are so breathtakingly beautiful it causes one to stare in awe at God's handiwork. We ate dinner at Black-Eyed Pea; yummy!

Weather: sunny and warm
Temp: 70's

May 11, Tuesday-Day 92 *Miles walked: 27.95*

Surprise! Surprise! Surprise! Dewey got up early to walk and I picked him up around 11:30 so we could get back to motel to meet Jim Smith. Dewey walked 27.95 miles today, a total of 2131+ miles. Jim already was at the motel when we arrived. We greeted Jim and then what to our wondering eyes should appear but Jim and Beulah Lykins who came to surprise us, and surprise us they did! Dewey and I both were speechless! Jim and Jim and Beulah flew out on the same plane into Denver. Jim Smith knew where we were but Jim and Beulah didn't, so, Jim Smith was able to be the guiding light for them. Jim and Beulah also treated us to dinner at Black-Eyed Pea. Friends are a wonderful gift from God. We have been blessed with the best. Jim Smith brought cards from our church family, some with dollars tucked inside. Also birthday wishes for me in card form. I also received a call from Betty Jaeger today with belated birthday wishes. What a day!

Weather: sunny and mild
Temp: 70's & 80's

May 12, Wednesday-Day 93 *Miles walked: 8.53*

Dewey got up not feeling great, blood pressure low. Jim Smith drove him up on 14 so he could begin walking. Jim stuck close because of Dewey's low BP. He walked 8.53 miles before returning at noon. They joined Jim and Beulah Lykins and me at The Egg and I restaurant. Jim and Beulah paid again. The Lykins had to leave right after lunch to drive to Denver to catch a plane back to Kentucky. Jim Smith also went to airport in Denver to pick up Diane. Dewey rested in afternoon while I went to hair salon. It rained all afternoon. We went to Black-Eyed Pea for dinner. Their special was chicken pot pie. It was good. We returned to motel (Quality Inn and Suites) and Dewey worked on computer and I read paper. A reporter interviewed Dewey for local paper. It may or may not get in. Dewey feels better tonight. Altitude must be affecting the BP. God continues to bless us every day.

Weather: cloudy then rain in PM
Temp: 50's

May 13, Thursday-Day 94 *Miles walked: 16.84*

We got on the road early. We had to pack car (in snow flurries). We drove up 14 thru Roosevelt National Forest and the Canyon to Cameron Pass. We had extraordinary weather, snow then sun, snow then sun. This went on much of the day. We stopped in Walden for lunch and there was a hail/sleet storm going on while we ate lunch in a neat café in Walden. (Dewey had Lemon Pecan Pie.) Jim and Diane Smith seem to be enjoying the Rockies. The Canyon looks totally different in snowy conditions but still awesomely beautiful. We arrived in Steamboat Springs, Colorado around 3:00. Dewey walked into Steamboat Springs, Colorado from 10 miles out. Dewey walked 16.84 miles today. We ate dinner at Boomerang's where you cook your own meat. When we got back to the hotel (lodge), a huge hail squall bombarded the area and that has been the story all day. God gave us safety and the handiwork was magnificent.

Weather: dry, snowy-then hail on & on
Temp: 30's to 50's

May 14, Friday-Day 95 *Miles walked: 17.59*

Today was an easy day for me. Jim and Diane did the back and forth driving for Dewey. They backed up past Walden, Colorado into the Mountains. The altitude was hard on Dewey today. He walked 17.59 miles. They saw wild life today which is always exciting. I ate a lite lunch at Freshies in Steamboat Springs. They took lunch and ate along the route. For dinner we ate at Smoke House Restaurant. The church took us out for Friday night date. The waitress was a neat gal. She was so interested in Dewey's walk, she gave us free dessert (yummy peach cobbler with ice cream — HUGE)! Each couple shared as the amount was too much for one. I even broke over and ate some bites. We have enjoyed having Jim and Diane share some time with us on the journey. Getting to know such nice folks better has been delightful. They will leave early (4:00AM) to go to Denver Airport to fly home. May God bless them and us as we continue our incredible journey.

Weather: sunny and cool
Temp: 40's & 50's

May 15, Saturday-Day 96 *Miles walked: 30.26*

The Smiths left at 4:00AM. We got up early but not that early! We had to go back past Walden to begin the day. We had a late breakfast in Walden at River Rock Café and Dewey continued walking. I shopped a little in Walden and wrote cards for a while. I took Dewey some lunch (bag lunch) and he rested a while before taking off again. He walked 30.26 miles today. We will return to 14 and he will walk there and on 40 to Steamboat Springs. We attended a 5:30PM service at United Methodist Church Steamboat Springs. Interesting service led totally by lay folks. Music was more praise music and service rather up beat as well. We grabbed a sandwich at the Villager Restaurant (reminded me of the old Howard Johnson-minus ice cream variety). Darlene called today and we got a nice e-mail from Barry Green. It was another lovely day. God continues to bless us with sunshine and blue skies. We are so grateful.

Weather: sunny and mild
Temp: 50's and 60's

May 16, Sunday-Day 97 *Miles walked: 32.77*

Since we went to church last night, Dewey got up early and so did I. He started walking early and walked 32.77 miles today. I followed him closely because he was walking in higher elevations. We took a break for lunch and ate at The Egg and I. Back to the grind! Later in day, we returned to Inn at Steamboat and Dewey took a hot bath and rested some before we went to dinner. We ate at Old Town Pub, food was good but a bit pricey. Dewey wrote some cards this evening and so did I. I worked on my deliberations for the newsletter. I signed up to drive for this trek, not to write musings, but I do anyway and makes me think which I guess doesn't hurt me. We will be in Steamboat Springs until Tuesday AM then move on to Craig, Colorado. After that, the pickins become slim for lodging. Talked to my boys today. They wanted to know if we were at our house. Went shopping and bought a few gifts today. God is very present with us. We continue to praise Him for His guidance and watchcare.

Weather: sunny to partly cloudy
Temp: 50's

May 17, Monday-Day 98 *Miles walked 32:18*

Dewey ate early and started walking from motel. I got up around 7:15 and got ready and went to breakfast in motel. I went by Visitor's Bureau to see if package from Jeana had arrived. It had not. I then went to PO to mail the shirts to Ann and Hester, Ruth & Bill, Maybelle & Darlene, DJ and Brennan. Then I went to gift shop to find something for Kim and Craig. I got them a cutting board with grape motif and a sun catcher with same theme. The store will pack and send the items. I then went to find Dewey so we could eat lunch. We ate at Lucille's. After lunch, we returned to walking spot and Dewey continued walking. I went back to motel to rest a while. I picked Dewey up around 3:30 to 4:00. He walked 32.18 miles today. Tomorrow we leave Steamboat Springs and The Inn at Steamboat. We've enjoyed our time here ($44.00 a night!) We will stay 1 or 2 nights in Craig, Colorado. We are so grateful that God continues to bless our effort.

Weather: partly cloudy
Temp: 50's & 60's

May 18, Tuesday-Day 99 *Miles walked 32:69*

HAPPY BIRTHDAY KIM

After an early call to arise, we packed the car, ate breakfast, checked out (I forgot to leave keys) and headed west. Dewey walked 32.69 miles today and we are staying in Craig, Colorado at a Holiday Inn (sp. rate of $59.00). Dewey walked all morning stopping only a few minutes to grab a sandwich and continue on. I picked him up in late afternoon and he got his hair cut and girl said to put cost of hair cut on walk. We put $20.00 in the fund. We returned to motel in time for Dewey to be interviewed by local paper. That took about an hour. We ate dinner in the restaurant in motel. Food was good and reasonable. Dewey got our reservations tonight for San Francisco, good rates at Comfort Suites at Airport. Tim, Matt and a friend are flying to Salt Lake City and joining us for Memorial weekend in Provo, Utah. Another beautiful day but windy in the afternoon. We thank God again for keeping us healthy, safe and providing good weather for our journey.

Weather: sunny and mild
Temp: 60's & 70's

May 19, Wednesday-Day 100 *Miles walked 32.22*

We woke up to our faces appearing in the Craig Daily Press in living color! The picture is good of both of us and the article by Jeremy Browning was one of the best that has been written re the walk. I had fun today sending copies to family and friends. I also got my hair done, looks pretty good. Dewey walked past Maybell today and we ate lunch at the Maybell Restaurant. Tomorrow we head for Dinosaur, Colorado. Dewey walked 32.22 miles today. He is doing better since altitude is not as high as it was in the mountains. We ate dinner in the motel again. Food is good and it is quiet. Also we don't have to go out looking for a restaurant. We made reservations for Provo, Utah for Memorial weekend at Holiday Inn. Sandy got us 3 free nights at a Day's Inn in Vernal, Utah. We are so blessed. Thank you God.

Weather: sunny and mild
Temp: 60's & 70's

May 20, Thursday-Day 101 *Miles walked: 33.72*

Another beautiful day to add to our repertoire of nice weather. God has truly blessed us with unbelievable weather. We reached the Utah border today and the lay of the land already looks different. Dewey walked 33.72 miles today as he walked to the border since he was so close when he got to Dinosaur, Colorado where we are staying tonight at the Terrace Motel. It is a mom and pop motel but it is clean. We will travel on to Vernal, Utah tomorrow and stay 3 nights at Days Inn. Sandy Clift got us 3 free nights there. We ate lunch and dinner at Miner's Restaurant, a mom and pop establishment next to motel. Food good and atmosphere typical of "open country" décor. We are going to bed early as there is nothing to do in Dinosaur. We met some nice folks today. One man gave $7.00 to the cause and the lady in gas station gave me a free coke for Dewey (her parents run motel where we are staying tonight). Blessings continue to mount! Thank you God.

Weather: fair & cool to mild
Temp: 60's & 70's

May 21, Friday-Day 102 *Miles walked: 30.91*

Morning broke early for us. We ate breakfast at Miner's and met the owner's grandson, Randy. He is graduating from High School next week and joining the Navy. I felt led to give him $20.00 (I think my mother was nudging me as that is what she would have done). Utah has a unique beauty. Every state does but Utah is such a contrast to Colorado with high desert plains on one side, big boulders on the other and snow capped Mountains in the distance. God has painted a magnificent picture. Dewey walked 30.91 miles today from Dinosaur to Vernal, Utah. We are staying tonight at Day's Inn in Vernal. Accommodations are complimentary for the two nights. On Sunday we will leave for Roosevelt. Dewey will be there by the end of the day on Saturday. We met interesting folks today. One lady stopped to see if I were okay since I was stopped with blinkers going waiting for Dewey. God's love continues to surround us.

Weather: fair and mild
Temp: 60's & 70's

May 22, Saturday-Day 103 *Miles walked: 29.84*

Dewey got an early start walking from hotel. He walked hard all day with a quick lunch break. He walked 29+ miles today from Vernal to Roosevelt. He will take tomorrow off to rest his feet. We will move ahead to Duchasne for the night. On Monday we will back track to Roosevelt for Dewey to begin walking. We sent two boxes home today with things we won't need any more til we get back. The wind was strong today and it makes walking more difficult. I followed Dewey closely today because of the vastness and wide open spaces. We ate dinner at Country Cafe. Food was okay. We are going to bed early so we can be gone by 8:00 in the AM. We had another nice day today. A few drops of rain fell, but not for long. God gives us lovely weather, safety and good health day after day. We are grateful for His blessings each day.

Weather: partly cloudy, mild (windy)
Temp: 60's

May 23, Sunday-Day 104 *Miles walked: Off day*

We packed the car and left Vernal early. We wanted to drive ahead and find a church in which to worship. We drove to Duchasne because there was a Baptist Church there. There are no Methodist churches in this part of Utah, all LDS (Mormon) churches. We looked and looked and finally found the little church. It was small (18 in attendance including us). A man (Larry) gave his testimony. It was moving! Drugs, alcohol etc. etc. had filled his being all his life, prison, broken relationships etc. God spoke thru him and no eyes were dry. They gave Dewey a chance to speak as well. Very caring people. We went to lunch with a couple who treated us to lunch. We went back this evening to their Bible study. Dewey did not walk today. We are staying in a "basic motel" tonight, clean but basic. Isn't God the one who guides? We know he led us to this place today. God doesn't make mistakes. He wanted us to share in Larry's testimony. Thank you God.

Weather: cool and sunny
Temp: 60's

May 24, Monday-Day 105 *Miles walked: 31.73*

Our day began at 5:20AM as Dewey wanted to call Jack Pattie (DJ Lexington) and give him an update on the Walk. We ate breakfast at the Country Grill and Dewey began walking before 7:00AM. He walked hard today, walking 31+ miles up hills, hills, and hills. We are staying tonight at Daniels Summit Lodge. It is really nice; fireplace in the room, large beds and big bathtub. Dewey really liked that. There is a restaurant on the premises and we ate there tonight. We will probably stay one more night here and then move on to Heber City for Wednesday night. It is only 14 miles from where we are tonight. However, Dewey is still a ways from Daniels and must back track tomorrow. When we arrived at Daniels Summit, there was snow (not on roads) but we did not encounter anything but clear, cool weather. Our guardian angel continues to intercede and God takes care of us. How blessed we are.

Weather: clear and cool
Temp: 50's

May 25, Tuesday-Day 106 *Miles walked: 32.59*

Dewey got up before 6:00 so he could begin walking a little after 6:00. I got up shortly afterwards. We fixed hot water in room for tea. Dewey walked 6+ miles before breakfast and then I picked him up so we could eat. I then returned him to starting point for the day and he walked 32.59 miles today. He will get to Heber City by tomorrow. We will go on to

Provo for 5 days and just go back and forth during those days. I saw a terrible accident today. I offered to call 911 but call had already been made. We ate dinner tonight in a restaurant in Heber City called Claim Jumper Steak House. We thought it was like one in Reno, Nevada but it wasn't that chain. It was okay but expensive. Not much choice either. The church keeps on giving! They gave Dewey a massage at the Lodge in their spa. It made Dewey feel better. Sandy Clift set it up and called to let me know. I don't like massages so I declined the offer. Blessings never cease. Weather continues to be in God's hands and we are grateful.

Weather: cloudy to partly cloudy and cool
Temp: 40's & 50's

May 26, Wednesday-Day 107 *Miles walked: 35.25*

Dewey began walking a little after 6:30 this morning. He walked 35.25 miles today thru Heber City. The walk was a lovely one thru beautiful canyons. We enjoyed our 2 nights at Daniels Summit Lodge. We are staying the next 5 nights in Provo, Utah (home of Brigham Young University). We grabbed a bite of lunch at Arby's in Heber City. I bought fruit at a grocery store. I also wrote lots of thank you cards. That's a never ending task. Heard from Dorothy Cummins today. Sue also called to tell us LaVerne broke her hip. She will have surgery tomorrow. Please Lord take care of her. We ate dinner tonight at Ottavio's in downtown Provo. We really enjoyed eating there again. Food delicious and atmosphere very Italian! We are staying at Holiday Inn and it is nice. The guest laundry is fairly close to our room and I did a load today. These chores are necessary and I get to do them; part of my "tour of duty". God gave us another beautiful day and we sing His praises for all His manifold blessings to us.

Weather: sunny and cool then mild
Temp: 30's and 50's

May 27, Thursday-Day 108 *Miles walked: 32.43*

We got up and had a delicious hot breakfast in the motel (Holiday Inn). Dewey started walking early up in the Canyon between Heber City and Provo. He walked 32.43 miles all the way thru Provo. We had another nice day weather wise. It went from sunny to cloudy and a few rain drops back to sunny. We saw on news that several inches of snow are expected tomorrow at Daniels Summit where we were on Tuesday and Wednesday. Once again, the weather holds out for us. God you are truly the "miracle worker". I had my hair done today at Penney's Salon. It looks pretty good. At least it is clean. We ate dinner tonight at Fridays. We ordered onion rings and we got a huge serving! Together we couldn't down them all. We hadn't eaten at Fridays the whole trip. We also went to Sam's for water and batteries. Thank you God for continually watching over us. Our journey is in your hands.

Weather: sunny, cloudy, sprinkles then sunny again
Temp: 50's

May 28, Friday-Day 109 *Miles walked: 32.36*

It was a good day for walking and Dewey amassed 32.36 miles today. I followed along as the route was unfamiliar, shoulders narrow and services sparse. We did find a neat little mom and pop establishment in which to eat lunch. It was The Family Tree in Santaquin, Utah (on highway 6). Everything is home made (bread, rolls, scones). The scones are the "signature item." They were 12 to 15 inches long and 3 to 6 inches wide in diameter as well. Dewey bought one, along with an apple sauce raisin loaf with cream cheese icing. He'll enjoy that with coffee. I wrote 16 cards today, my daily task. We want all contributors to know how much we appreciate their generosity and care re the walk. Tonight we ate at Cracker Barrel. It was Cracker Barrel good. We had another day without rain. God is blessing our journey in so many ways and giving us daily miracles. We are grateful for God's daily bounty.

Weather: partly cloudy (rain in late PM)
Temp: 50's – 60's

May 29, Saturday-Day 110 *Miles walked: 25.64*

I had an easy day today as Dewey walked locally for 25.64 miles as Tim, Matt and friend flew in to Salt Lake City from Cincinnati for the weekend and Dewey wanted to be close by. Tomorrow the guys will take turns walking with Dewey. Total miles =2600+. I won't drive tomorrow. It'll be my day off. Dewey and I will go to church at the Episcopal Church at 8:30 in the AM. There are very few churches in the area other than Mormon. We ate dinner tonight at Ottavio's (Tim paid). Then Dewey treated us to Yogurt (they had sugar free) at Hogi Yogi. The weather was great today. After a cloudy early morning, the sun came out and the day was lovely, a little on the cool side. Bad weather seems to be all around us, but the Red Sea parts and we are on dry land. With God, all things are possible. Thank you God, yet again.

Weather: sunny and cool
Temp: 50's

May 30, Sunday-Day 111 *Miles walked: 12.33*

Today was another beautiful day in God's world. Dewey and I attended an 8:30 service at St. Mary's Episcopal Church in downtown Provo. It is the only thriving protestant church in the area. All the other churches are Mormon. When we got back from church (woman priest conducted service — 5 in attendance, their later service had many more) Dewey, Tim, Matt and Vaughn walked for 12.33 miles. I had a respite from driving today so I relaxed, wrote some cards, took a nap, and watched a little TV. I also dropped the cards off at the P.O. This evening we took the guys out for dinner at The Family Tree in Santaquin, Utah, (15 miles from Provo.) The guys were shocked by the scones. They are really BIG! Bad weather is behind us and all around us but our road is dry. God has truly taken care of us — answering prayers from all directions for us. Thank you God for your watchcare and all the prayer warriors from sea to shinning sea.

Weather: sunny & cool
Temp: 40's to 60's

May 31, Monday-Day 112 *Miles walked: 32.87*

MEMORIAL DAY

What gorgeous weather we are having! Sunshine, dry roads and even a breeze or two. It was warm for Dewey as he walked today but he still clocked 32.87 miles. Tim, Matt and Vaughn got away early for Salt Lake City to fly to Cincinnati. We left to head west on highway 6 and 50. We are staying tonight at Delta, Utah (also tomorrow). We got yogurt this PM at Hogi Yogi! It's good. Dewey is way past Delta but places to stay are few and far between. On Wednesday night we will stay in Border, Nevada for one night before trekking on to Ely, Nevada. The route is high desert. I followed Dewey closely today, taking him cokes and checking on him. Services are pretty much nil across the route. Dottie South called today. I also talked to LaVerne. Thank you God that she's doing well. Thinking positively certainly helps her. As we continue this journey, I'm reminded of another favorite hymn of mine. "To God be the Glory — great things He has done". Thank you God.

Weather: sunny & mild
Temp: 50's to 70's

June 1, Tuesday-Day 113 *Miles walked: 35.19*

Dewey began walking early to "beat the heat". It was a hot day. The lay of the land is high desert; sand and desert plants. I followed closely so Dewey could have water or an icy coke and cool off for a few minutes every few miles. He walked 35.19 miles today. We are staying another night at Best Western in Delta, Utah. We ate an early supper at Top's café in Delta. We ate some Hogi Yogi Yogurt late and it was good. I had sugar-free peach and it was tasty. I wrote a lot of thank you cards today to contributors to the Walk. I read a little and did a lot of back and forth driving. We had another beautiful sunshiny day. God continues to bless our effort on this venture. It is hard to take everything in and I'm in awe of how each blessing has been followed by another blessing.

Weather: sunny and warm
Temp: 50's to 90's

June 2, Wednesday-Day 114 *Miles walked: 38.03*

Today was early for us and without a lot of sleep. Folks partied outside our room until wee hours of morning. I called "desk" and they said they would take care of it. I'm not sure they did as noise continued until after midnight. I intend to let AAA know what this motel (Best Western in Delta, Utah) allows! Our day went well. Dewey walked and walked and I stayed close by because of heat and lonely road. One lady told me the 80+ miles was the loneliest road in America. I'm sure it is in Utah. For over 80 miles there are no services of any kind only road and desolate country side. Dewey walked 38.03 miles today. We are staying the night at Borden Inn right on the Nevada border…basic motel but clean. Restaurant attached and food okay. We will continue toward Ely, Nevada tomorrow and stay 2 or 3 nights there. The weather continues to be gloriously sunny, but hot. God is a very present help in this journey. We are grateful for each new day of blessings in abundance.

Weather: sunny and hot
Temp: 70's – 90's

June 3, Thursday-Day 115 *Miles walked: 34.54*

We are grateful to be in the next to last state in our unbelievable journey (Nevada). The final state of California is a lot closer than it has been (30 more days, I think)! Got my hair done today, looks okay. I had a first today. I got stopped by police outside Ely, Nevada for speeding. I was going 70+ in a 55 miles zone, but I hadn't seen the speed zone. 70 is legal on roads, but we were in closer to town. Officer was very nice and because of our mission only warned me. I was upset with myself for the infringement but grateful I didn't get a ticket; another blessing from above. Dewey walked up, up, up today and he was tired. He started walking 6:30AM (PT) while it was cooler. By midday it was in the 90's. That's draining! We are staying at a Best Western in Ely, Nevada. Sandy got us a complimentary night here and we will stay 2 more nights. Weather continues to "out do itself". Aren't God's gifts to us treasures to behold! We ate dinner tonight in the Jailhouse Coffee Shop. It was tasty. Dewey walked 34.54 miles today and will try to repeat it tomorrow. Thank you God for everything.

Weather: hot, dry and sunny
Temp: 50's to 90's

June 4, Friday-Day 116 *Miles walked: 32.58*

We arose at 5:30AM (PT). Dewey's partner, Tim McCann, called thinking it was 6:30AM not realizing we had entered Pacific Time Zone. Dewey razzed him a bit, but the fact of the matter is we were already up. Dewey struggled today with low BP again but still managed to clock 32.58 miles. We got lunch at the Red Apple Restaurant. Then we had dinner at Evah's Italian eatery. Food okay. Mark Streety called me today and hopes to join Dewey for a couple of days to encourage him and walk a day. I gave him some possible dates and he will try to work something out. It would be a nice "surprise treat" for Dewey. It was very hot today, high 80's but the sun was shining and all's right in God's world. I've been prone to take so many things for granted, but my eyes and ears continue to be opened daily to the wonderful world God has created…and I'm grateful to be a small part of His creation.

Weather: hot, dry and sunny
Temp: high 80's

June 5, Saturday-Day 117 *Miles walked: 35.61*

Up, up and away! That was Dewey's walking for today all the way up to Robinson Summit. I played "water and coke girl". I also shopped for bananas and donuts for Dewey to snack on during rest breaks. Another sunny but hot day. We thought it might get colder as we climbed "Jacob's ladder" toward heaven but it didn't. We have given out of places to eat in Ely, not that there are that many. We did return to the Jailhouse Coffee shop for dinner tonight. Neither Dewey nor I were that hungry. We are going to bed early as Dewey plans to walk around 6:00AM then we plan to go to Ely First United Methodist Church in the morning. It meets in a Masonic Lodge while building a new church. We aren't sure where the old building is. We are most fortunate folks to have so many people interested in the walk and a God so generous in granting blessings. Thank you seems inadequate.

Weather: sunny, dry and hot
Temp: high in 80's

June 6, Sunday-Day 118 *Miles walked 38.06*

Neither Dewey nor I slept very well last night. Don't know why. We woke up around 5:00 or 5:30 so Dewey decided to walk early; eat breakfast, pack the car, check out; go to church (9:30AM) in that order. The little church we attended meets in a Masonic Lodge right on Main Street. It was a meaningful service with their church family. We were a part of the service by taking Communion with them. There were 20 including us. They will be praying for us as are other churches all along the way. One man gave Dewey $20.00 for the cause. I locked the keys in the car and we now know that ON STAR works. We were in the car in less than 10 minutes. I was upset to be so foolish, but my guardian angel made sure everything worked out ok. "God is a very present help in trouble." Dewey walked hard today for 38.06 miles. We must push hard everyday across the desert. What else is there to do! Ruth called today; great to hear her voice; also talked to Judy (Atlanta).

Weather: sunny, dry and hot
Temp: high in 90's

June 7, Monday-Day 119 *Miles walked: 38.26*

We stayed in Eureka, Nevada at a Best Western. It is a nice motel, recently remodeled. They served a great continental breakfast including fresh fruit (strawberries, melons of all kind). We both slept well last night, going to bed at 8:00PM (PT) and arising at a little after 5:00AM by Hannibal and his elephants moving above us! Those folks were really heavy footed! Dewey started walking a little after 6:00AM and I played handy person again today; getting cokes, chips etc. Anyone who says the desert is flat hasn't been across the route we are taking because Dewey climbed again today; all day. Wind was a factor too. Theresa called to tell me Tim and Tom are coming to walk the bridge. El called and she and Don are coming also. The list is mounting for San Francisco! What an exciting time we will have! Another beautiful day. Dewey walked 38.26 miles today. We thank you God for keeping us in the palm of your hand. It is no accident that our journey has been in His care since the beginning.

Weather: fair & cool to warm
Temp: high 80's

June 8, Tuesday-Day 120 — *Miles walked: 38.18*

Another night at Best Western in Eureka, Nevada. We both slept okay last night but we got up early again (around 5:20AM) Dewey starting walking early, a little after 6:00AM and walked for 38.18 miles with the lay of the land looking the same for all 38 miles. Dewey's total miles are 2991+. Dewey is within 2+ miles East of Austin, Nevada. We will move ahead to Fallon, Nevada and stay 2 or 3 nights. We will need to back track but there is no place to stay in Austin or points in between. We've had phone calls from Helen Fassler, the church group and Dewey talked to Harold Henson, Bob Ebersole, Newton Sanders and Jim Wilbers. I did laundry today so we are okay for another few days. Weather was colder today for Dewey to walk. I stayed close by because there were no services at all. I took cokes, chips, etc for us to munch on. We had to find bushes for our bathroom needs. We count our blessings every day and are grateful for each new day's experience along this walk.

Weather: sunny and cool
Temp: 40's to 60's

June 9, Wednesday, Day 121 — *Miles walked: 32.67*

ANOTHER DAY TO REMEMBER

Jeepers! We started out in beautiful sunshine bright and early this morning and what do you know? When we reached the summit which leads into the town of Austin, Nevada it started snowing! And it snowed and snowed. When we reached the starting point for Dewey to begin walking it was still snowing but didn't seem hazardous so Dewey walked thru Austin and on towards Fallon where the weather had cleared some so Dewey was fine. God has given us yet another sign that He controls the elements and His world. How grateful we are to be given the privilege of experiencing so much of God's world. The variety has been awesome! We are staying tonight at a Best Western in Fallon, Nevada. We thought Dewey would have high-speed internet but it wouldn't work. Oh the perils of interneting! We ate dinner at Cock n' Bull and the food was reasonable and good. It will be early to bed as usual because 5:00 comes sooner than later. Dewey walked 32.67 miles today for a total of 3023+ miles.

Weather: sun, rain, snow, sun again
Temp: 30's to 50's

June 10, Thursday-Day 122 *Miles walked: 35.8*

We got up early and Dewey had several call backs to make. One was to United Methodist Communications out of New York. They are going to interview him for an on-line program re his walk. This will be done at a studio in Reno on next Wednesday at 11:00AM. He started walking after the call backs and walked a total of 35.8 miles today, total miles 3058+. Some of the walking was on a levee and it was difficult to find places to pull over along there. We went out to get a closer look at Sand Mountain; took pictures there. We grabbed lunch at Heidi's and ate dinner again at Cock n' Bull Restaurant. Dinners are reasonable there and good. Talked with Dorothy Cummins, Sue Daniels and Martha today. I will call LaVerne Nelson tomorrow. She is home from the hospital now. It was partly cloudy today and temp was cool to mild. How grateful we are that God continues to bless us mightily.

Weather: partly cloudy and mild
Temp: 40's to 60's

June 11, Friday-Day 123 *Miles walked: 24.65*

Our day began early as usual. Dewey was not feeling well today and as a result he only walked 24.65 miles. We left Fallon this AM and headed toward Carson City (Capital of Nevada). We are staying two nights at a Best Western in Carson City. Nevada is a beautiful state. God has given

Nevada its own personality just as he has all the states we have been through. There is real variety in God's handiwork. Sandy Clift called today. I also talked with Laverne Nelson. She is home now and recuperating from hip surgery. Weather was good today. We have been so blessed in every aspect of this amazing journey. Ann and Hester will arrive on Sunday for a few days with us. All of us will be staying in Reno and will go back and forth re Dewey's walk.

Weather: sunny and mild
Temp: 60's

June 12, Saturday-Day 124 *Miles walked: Off day*

Dewey was unable to walk today. He started feeling bad yesterday but still walked. However, he was too sick to walk today. He had a bad case of diarrhea and general feeling of weakness. I played Dr. Mom or Nurse Dotty. I hope I don't get it but I wouldn't be surprised if I do. I've eaten in hotel for meals today. Food is not expensive and it is good. Dewey has only drunk 7up and eaten crackers. I also brought him some hot tea with lemon & honey. Dr. Ellis (who delivered Craig) used tea with a little lemon for every cure all when I was pregnant. We leave in the AM for Reno. Ann and Hester will arrive for a few days with us. If Dewey feels like it we will go to church at 8:00 at Methodist Church here in Carson City. I am grateful that Dewey is feeling better. I believe God takes care of us one way or the other.

Weather: sunny
Temp: 70's to 80's

June 13, Sunday-Day 125 *Miles walked: 30.02*

Today we got up early. Dewey walked over 30 miles today. We went to an 8:05 service at First United Methodist Church in Carson City. A clergy couple pastors the church. Reverend Dixie spoke this AM. Message meaningful and church was interested in the walk. All across the USA Methodists are praying for us and we definitely feel every prayer. We checked out the walk over Sierra Nevadas thru Lake Tahoe. It is a gorgeous drive and Dewey will definitely enjoy walking it. Ann and Hester arrived today. They got to Reno before we did. We all checked in around 4:00PM. We drove around the city and ate dinner at Claim Jumper Restaurant, food huge and delicious. Dewey ordered a Mother Lode (Chocolate Cake) for Kim. She needs a lift right now and she'll love the cake. We have a lovely room on the 14th floor of Atlantis Hotel in Reno. Ann and Hester will drive (walk) tomorrow and I'll get a break. Actually, I'm going to try to get a perm. I've had too many bad hair days to suit me. God has again blessed our effort. Dewey is feeling better and the rest did him good.

Weather: fair and mild
Temp: 80's

June 14, Monday-Day 126 *Miles walked: 21.12*

RENO

Dewey, Ann and Hester left bright and early for Lake Tahoe and the 5% grade-hill that was 10 miles long! Dewey also walked another 10+miles for over 20+ miles today. Ann & Hester went to Ponderosa and Virginia City today. I stayed at hotel to rest and get a perm. I neglected to tell the stylist I wanted a curly perm so I think I only got body wave; $80.00 for body! Oh well, at least it looks better. Just hope it stays for 5 more weeks. Dewey will get a perm on Wednesday AM (8:00AM) and I warned him to ask for a curly perm. We went to dinner at Olive Garden and then Dewey, Ann and Hester did our laundry, another break for me. I plan to rest again tomorrow and maybe eat a late breakfast in hotel. I feel indebted to the many prayer warriors who have lifted us up daily. God has answered their prayers; for life for us has been exceedingly rewarding.

Weather: sunny and mild
Temp: 70's and 80's

June 15, Tuesday-Day 127 *Miles walked: 23.27*

HAPPY ANNIVERSARY, KIM AND CRAIG

Today was another "free" day for me. Ann and Hester drove Dewey and they were able to shop around in Carson City, Nevada. Dewey walked 23.27 miles today and walked into California, the last state of the walk! Hooray! I read, wrote my deliberation, scheduled the car for service etc, etc. I had a nice lunch in the Purple Parrot in the Atlantis Hotel. Food good, service only fair. We checked out where Chevy dealership is for service at 8:00 in the AM. We went to TGI Fridays for dinner. I got fish and shrimp, good. We also had an order of onion rings. We came back to hotel as Dewey had to retrieve a Fax re interview at 11:00 tomorrow. It is hard to put into words just how much this journey has enriched our lives. God has been the guiding force and the Constant throughout. We have had good weather, safety and good health. God has given us many miracles to treasure.

Weather: sunny and mild
Temp: 80's

June 16, Wednesday-Day 128 *Miles walked: Off day*

It has been "hump day" once again. Dewey took the day off and got a much needed perm. I took the car to be serviced at 8:00. Dewey had an interview for United Methodist Church at 11:00. Ann, Hester and I took him to place for interview and we went to Heidi's for breakfast. Afterwards, we went to Claim Jumper to pick up cake to send to Kim and Craig. Will they be surprised when they see it! (All over $100.00 worth) We then drove down to Truckee, California, a quaint town near Lake Tahoe. From there we went to Donner Pass Park. We came back thru Lake Tahoe where we all had "Ivory ashes" to scatter in the lake. And what do you know? Swimming in the lake was a Golden Retriever that looked very much like Ivory girl (even though Ivory was a yellow Lab). It was almost as if Ivory were wooing "all is well". God granted us yet another blessing. Thank you again. Dewey also had an interview with radio station in Sacramento. We ate dinner once again at Claim Jumper.

Weather: sunny and mild
Temp: 80's

June 17, Thursday-Day 129 *Miles walked: 30.81*

Our day began early as we had to pack car etc. before "getting out of Dodge". Ann and Hester met us by the coffee bar. They rented a car for today and are going to Virginia City again. We, on the other hand, were "off to the races" (walking, of course). Dewey walked around 30.81 miles today and will have to push on for the next two days to get closer to Lodi, California where we are staying until Monday. I will have to back track because there are no very good places to stay otherwise. We are tired tonight after going up and down the hills today. Kim got her cake and couldn't believe how BIG it was. We ate dinner tonight at Janet's Richmaid Restaurant in Lodi. Dewey is trying to get online in order to send Newsletter. He is having a dog of a time having any success! Another beautiful day to add to all the other ones we've consistently enjoyed day after day. Our God is an awesome God indeed. Thank you Lord for being a shining light day after day.

Weather: sunny and mild
Temp: 80's

June 18, Friday-Day 130 *Miles walked: 31.44*

We got "on the road" early this AM after a quickie continental breakfast in the motel. We are staying at a Best Western in Lodi, California. Dewey walked 31.44 miles today and we had to go back close to Kit Carson, California for him to begin walking. He was interviewed by a local paper today and also received a call re an article on him for the Interpreter Magazine (a Methodist publication). We ate dinner tonight in Clements, California at Lucas Steakhouse Restaurant. It was an attractive place and food was good. We got back to motel around 7:00 (PT) and Dewey worked on computer and I watched a little TV. The weather was lovely again today, high in 80's. I stopped at a shop today and the lady was interested in the walk. She gave me bottled water for both of us. It was another day of lovely weather and meeting local folks. God certainly knows how to give us a "perk" just when one is needed.

Weather: sunny and mild
Temp: 80's

June 19, Saturday-Day 131 *Miles walked: 27.53*

It is hard to believe but another week has come to an end. One week just moves into another one so quickly. We have been on the road 19 weeks next Tuesday and the end is in sight. Dewey has walked 3244+ miles to date. He walked 27.53 today and is within a few miles of Lodi, California. We drove around Lodi this evening after we ate dinner at the Golden Ox Family Restaurant. Food was okay, not pricey. Dewey shopped at Ross. He likes their socks, so he stocked up. We talked to El, Chuck, Jim W., Bill, Ann & Craig. Chuck has been elected President of the KWC Alumni Board. We are proud of him (it is a 6 year commitment). As I reflect upon this journey, I am deeply aware of the important part our friends have played in the success of our efforts. Their love, prayers, contributions, and interest have given us the encouragement we have needed to continue to the end, the Golden Gate Bridge. Thank you God for the gift of friends.

Weather: sunny and mild
Temp: 60's to 80's

June 20, Sunday-Day 132 *Miles walked: 7.62*

Our day started early and was a busy one. We went to an 8:00AM service at First United Methodist Church in Lodi, California. The minister was South African, accent and all (been in US 20 years). Loving and hospitable church. We left after church and drove what we thought would be Dewey's route to San Francisco, but we encountered problems re roads becoming freeway and/or Interstate. Dewey will need to check with AAA tomorrow about alternate routes. We drove on to San Francisco and visited Fisherman's Wharf and other spots of interest. We drove across Golden Gate Bridge and folks were walking it. We hope it will be possible on July 4th as it was today. We got in a traffic jam on highway 12 on return trip. I thought we would never get back to Lodi. We stopped at an Applebee's on edge of town to eat. Dewey was interviewed by a free lance writer out of Nashville for articles re walk. Thank you God for another day of blessings. We stand in awe of your goodness to us.

Weather: sunny and mild
Temp: 60's to 80's

June 21, Monday-Day 133 *Miles walked: 28.77*

Is it really possible we have been on the road 133 days? The answer is YES! Dewey walked 28.77 miles today and is within "shooting distance" of Napa Valley. Jim Wilbers will join us tomorrow to walk with Dewey for a few days. We have stayed in Lodi, California for 5 nights and will move on tomorrow to possibly Fairfield, California for the night. I did laundry today and Dewey got RX refilled at Wal-Mart. Folks there were interested in Walk. I worked on my deliberation for next newsletter. My next to last one. For the last one I'm going to write about my friend the Tahoe! The weather here in California has been "picture perfect" in every way. I wish I could describe the beautiful flowers that are everywhere. God's paintbrush has worked overtime out here! I am truly at a loss for words when I try to pen my feelings about God's goodness and His watchcare of these two slightly weary travelers.

Weather: sunny and mild
Temp: 60's to 90

June 22, Tuesday-Day 134 *Miles walked: 20.86*

Another day, another mile or two or three, actually 20+ for a total now of 3271+. Jim Wilbers arrived to walk with Dewey for several days. Dewey walked early before Jim arrived (5+ miles) and together they walked 15+ miles. It was a busy road (highway 12) with no shoulder in many places so they walked in brush much of the time. The wind was also strong impeding the walk somewhat. Jim parked his rental car in a shopping center and I took them back to walking spot. I picked them up after 15 miles and they will return to that spot tomorrow. We are staying at a Best Western in Fairfield, California. We will move on tomorrow to stay in Yountville, California at a train motel. It should be interesting. The weather today was beautiful and I can only say another thank you God for the blessings we have had all along the way. We drove the route (twisting and turning) which the guys will walk thru Napa and S. toward the ocean.

Weather: sunny and mild
Temp: 60's to 90's

June 23, Wednesday-Day 135 *Miles walked: 18.34*

We got up at 6:30 (actually I was already awake) and Dewey and Jim got ready first and went on to breakfast. I got ready while they had breakfast and Dewey brought me a glass of milk which is all I wanted. Jim left his car at the motel and I took them to the starting point to begin walking for the day. While they walked in AM, I got my hair done. Looks okay. Then I wrote some cards and read a bit. At 11:00, I went to Black Bear Diner to eat lunch. I got a Taco Salad. Dewey called as I was finishing and I went and picked them up for lunch. They returned to the walk. Jim walked 15 miles today and Dewey walked 18+ miles. We are staying tonight in the Railway Inn in Yountville, California. Yountville elevation is 105 feet. We had dinner at Marie Callender's Restaurant. The chicken pot pie was very delectable. They also featured a good variety of desserts but we were all too full to indulge. We enjoyed great and glorious weather again today. God is the author of the "Days of Our Lives." Each day is filled with blessings.

Weather: sunny and mild
Temp: 60's to 90's

June 24, Thursday-Day 136 *Miles walked: 15.53*

We packed the car and got away early. We stopped at Marie Callender's for breakfast. We called several motels re getting lodging for Thursday through Saturday but didn't have luck due to a NASCAR Race in area over the weekend. Soooo I drove back up to Yountville and got a room at Railway Inn again for next 3 nights. We are staying in the same room. Dewey and Jim walked 15+ miles today and are getting closer to San Francisco. Tomorrow the guys will walk for a couple of hours and then we are going to see Doug and Ellie Norris who are at a family camp. We will get to reconnect with their kids and meet their grandkids. It will be a special time for us. Dewey had to call Ann and dictate the newsletter to her because we don't have internet access where we are staying. We ate again at Marie Callender's. Dewey and Jim bought a whole banana cream pie for $5.99. Once again sunshine flooded our lives. God continues to smile on us. Amen!

Weather: sunny and mild
Temp: 60's to 80's

June 25, Friday-Day 137 *Miles walked: 7*

Today was a day of "mixed greens". We got up early for Dewey and Jim to walk several miles before leaving for Doug and Ellie's family camp. So after breakfast, we went to starting point of walk and the guys walked 7+ miles before leaving for camp. We had breakfast at Marie's before beginning the walking. We ate lightly at lunch in Lodi before heading south and east into the Mountains. It took us 4 hours to get to the camp (arriving around 2:30). We met all of Doug and Ellie's family (3 sons, 3 daughters-in-law and 7 granddaughters.) One is expecting again and they are hoping for a boy one last time, they say. We visited and then ate dinner at camp with them. Food very good, nicely prepared and 200+ people eating together, quite impressive. We have another memorable day to treasure and we are grateful once more to God for allowing us the opportunity to reconnect with some special friends from the past. We got back at 9:30 PM, tired but happy for the time spent with friends.

Weather: sunny and mild
Temp: 60's to high 80's

June 26, Saturday-Day 138 *Miles walked: 12.27*

My "inner" alarm clock buzzed at 5:30AM! We all got up so the guys could begin walking very early. We had to return to starting point some 20+ miles away. We didn't even stop for breakfast. Just grabbed juice (tomato) and coffee and donuts for guys. They walked thru Petaluma and out; unfortunately a man directed them incorrectly. They walked "out of the way" for 3 miles. They walked 12+ miles today. Visitor's Bureau got things straightened out and Dewey will head toward the ocean and highway! Tomorrow after we go to church, hopefully early, we will move on to another motel, but don't know where yet; maybe one on highway 1 along the ocean. Jim leaves in the morning to return to Fl. Dewey has enjoyed the companionship of someone besides me I'm sure and Jim walked over 60 miles with Dewey also. I shopped a little today but only bought a couple of things for Kim. We have enjoyed sunny California. We have certainly been touched by the beauty here. God's handiwork is evident everywhere we turn. I talked to Darlene, Betty Jaeger and Martha.

Weather: sunny and warm
Temp: 60's to 90's

June 27, Sunday-Day 139 *Miles walked: 15.69*

Jim left this morning and Dewey and I attended a 9:30 worship service at First United Methodist in Napa. The minister is a woman and extremely bright; enjoyed her sermon and the service. We then ate brunch at Marie Callender's. It was very good. We left the Napa Valley area and started south checking out routes for Dewey to get into San Francisco. He walked 15.69 miles today and hopes to reach his goal by Thursday. One of the routes we checked out was very curvy and hilly. Dewey won't take that one but one that is less hilly and curvy. Talked to Ruth and Bill today and Craig. We need to get facts re the B&B up in St. Helena's where we are booked for next Monday (July 5th). Kim is the one we need to talk to about this. It is wild how it all happened. Dewey called to say hello and maybe meet for coffee. Before we knew it, we were staying on Monday night! We are staying in Four Points Sheraton in San Rafael, California tonight. We almost gave out of gas today but our guardian angel took us to a gas station without incident. Thank you God for taking care of us yet again.

Weather: sunny and warm
Temp: 80's

June 28, Monday-Day 140 *Miles walked: 15.51*

(Is it possible) the journey's end is so near. Yes, yes, yes, Dewey has 15+ miles to go to get to the Bridge. He'll try for that tomorrow. He had to change his route because of so many hills (Mountains) right here at the end. He has to hook up with bike trail that leads into Sausalito tomorrow. That's right at the foot of the Golden Gate Bridge. We are staying tonight at a Best Western in Corte Madera, California (near 101) not too far from tomorrow's starting point. Dewey is ready for it all to be over and so am I. The journey has been incredible. God has been our Constant and friends our encouragers. But all good things need closure and the time has come for the final AMEN! We have had many calls today from friends all over. What a blessing that has been all along the way. We ate dinner tonight at Max's on the hotel grounds, tasty. Dewey is preparing the final newsletter before THE FINISH. Thanks be to God for holding us in the palm of His hand since February 10th.

Weather: sunny and warm
Temp: 70's & 80's

June 29, Tuesday-Day 141 *Miles walked: 10.46*

THE JOURNEY ENDS, UNOFFICIALLY

Today was a special day. Dewey walked to the bridge, Golden Gate, that is. I shouted Hallelujah…almost sang the Hallelujah chorus…a capella. It is hard to believe that the journey is over! We still have the celebration on Sunday at GGB with all those coming to share the special moment with us…special service at 9:30, walking the Bridge at 10:00 and on to the beach (aporox. 5 miles total). I will probably drive except I'll walk down to the beach when Dewey steps in the Pacific Ocean. We are visiting today and tonight with Doug and Ellie Norris, Dewey's friends from Seminary. We have kept in touch through the years and visited with each other a few times; great to reconnect again with these special friends. Many folks have called Dewey today to congratulate him on his accomplishment. There is somewhat of a let down after being "wired" for almost 5 months, but there is also a sense of relief in operation as well. This journey was always in God's hand and we thank Him for leading us every step of the way.

Weather: fair and mild
Temp: 70's

June 30, Wednesday-Day 142 *Miles walked: Off day*

We got up early and Doug fixed us a nice breakfast which we enjoyed at 8:00AM. Ellie fixed lunch and had invited another couple to come as well. We ate at 1:00. Ellie fixed pork loin and all the fixins. She is a great cook.

We enjoyed our visit with them and meeting their friends Nancy and Bob. Dewey and I left at 3:00 and arrived at Comfort Suites around 4:00. Fortunately the traffic was not too bad for us as we traveled west or south! We aren't too far from the airport and about 15 miles into San Francisco. We ate dinner at the Hungry Hunter and food was good. This was the restaurant we were supposed to have the dinner but cancelled it. We will go there again on Sunday with a group after the walk. Again, we had a nice day. The sun came out around 11:00 and pretty rest of day. We continue to count our many blessings. God has done great things for us.

Weather: fair and mild
Temp: 70's

July 1, Thursday-Day 143 *Miles walked: 4.04*

ANOTHER UNOFFICIAL END

The phone woke us this AM at 8:00. It was Dewey's friend, Bob Hutchinson from Mountains in Kentucky. He is giving $5,000.00 to the Walk. We figure that was a call worth waking up for! We had breakfast in motel and met a cute couple from Hawaii. She was excited about meeting a "celebrity". Wanted pictures with both of us. We left around 10:00 to go to GGB for Dewey to walk it and route to the beach. He did this and on down Lincoln Blvd to Baker Beach. We both went to beach and touched the Pacific Ocean. The walk is now complete. Dewey will repeat the walk on Sunday with the group if security allows. We went down into Sausalito in PM and ate at Winship Restaurant again. Our buddy Bill wasn't there today but restaurant staff knew about the Walk. We shopped a little and returned to motel. Went to dinner at Max's Opera Café. Darlene and Sandy got in around 7:00 and came to "surprise" us around 9:30. Great to see them. Thank you God for their safe flight and your watchcare of us as well.

Weather: cloudy (foggy) then sunny and cool
Temp: 60's to 70's

July 2, Friday-Day 144

Dewey, Darlene, Sandy and I went to Fisherman's Wharf (Pier 39) (10:00) and met Jeana, the Gibsons and the Hahns. We went shopping and then met for lunch at 12:30. We ate at Pier Café. Then Sandy and Darlene went to Ghirardelli Square and afterwards we took a ride on Lombard St.

(most crooked street in the world). We all had dinner together along with the Wethingtons, McCanns and Noyes (13 in all) at Hungry Hunter. Food was excellent. Dewey went to airport to pick up Ruth and Bill. Tim went with him. While they were gone, I had a first experience of the 144 days. The toilet overflowed and made a "first class" mess. I had to call maintenance. Isn't it something how I am always around for the "potty excitement"? Dewey has to go to GGB by 10:00 in the AM for pictures etc. It is wonderful sharing this special time with friends, God's special gift to us. Thank you, Lord. AMEN

Weather: sunny and mild
Temp: 70's

July 3, Saturday-Day 145

Our day began with a quickie continental breakfast in the hotel surrounded by our friends. What a joy to have so many join us for "the end of the story". Ruth, Bill, Ann, Darlene, Sandy, Dewey and I climbed into the silver Tahoe and we headed for Fisherman's Wharf to look and shop. It was a beautiful day, sunny, cool, mild and windy like it gets around the water. Dewey had to go with the film crew to plan the video re walk. He returned around 1:15 and we drove over to Sausalito to eat a late lunch at Winship Restaurant. Dewey was a hungry hound so we drove "lickety split" over the Golden Gate Bridge down into Sausalito. We enjoyed a great lunch/brunch and Dewey's was complimentary. After we left, Shannon, the hostess, came looking for Dewey because the busser found a money holder and low and behold it was Ann's! What an honest young man! We drove through Chinatown and on Lombard Street again. Tim and Tom took us to dinner at Marie Callender's. Yummy. We are grateful to God who has blessed us throughout our journey. AMEN.

Weather: gorgeous
Temp: 70's

July 4, Sunday-Day 146 *Miles walked: 4.04*

THE JOURNEY ENDS OFFICIALLY

Thirty-eight strong gathered at Vista Point on the northside of GGB to visit, worship and prepare for the final leg of the journey. TV coverage was there as were local papers. The documentary crew was there to film as well. It is impossible to believe that what began on Feb 10th has ended; the culmination of a life-long dream for Dewey. My thoughts are simple at this moment. God gave the dream and also the fulfillment of the dream. My prayers for safety, good health and good weather were all answered and the prayers lifted up by prayer warriors across our great nation were answered as well. Having friends gather with us from many places was a special gift from above. Our blessings have been rich and innumerable. What can I say except Thank You Lord. Your bounty has been poured out for us from day 1 to day 146. What a story!

The following were a part of the entourage crossing GGB: Sanders, Derringtons, Freys, McCanns, Noyes, Hahns, Gibsons, Ann, Darlene, Sandy, Crew (Video), Ryles-Smith, Nicole, Myles, Stonecyphers, Norrises, Jeana, Michael Fox, Jill, Jim McCord, Wethingtons.

What a march! What a Celebration!

Dewey's total mileage for the journey was 3355+ miles — Wow! Imagine That!

Bishop Beverly Shamana from California sent a proclamation concerning Dewey's walk. It was a wonderful surprise ending to the beach experience.

Weather: Foggy early then nice
Temp: 60's to 70's

Epilogue = Return to Kentucky

Day 147 Fairfield, CA — Best Western

Day 148 Yountville, CA — Railway Inn

Day 149 Reno, NV — Atlantis Hotel

Day 150 Ely, NV — Best Western

Day 151 Heber City, UT — Daniel Summit Lodge

Days 152-153 Steamboat Springs, CO — Inn at Steamboat

Day 154 Ft. Collins, CO — Best Western Kiva

Day 155 Abilene, KS — Holiday Inn Express

Day 156 Springfield, IL — Hampton Inn

Day 157 Indianapolis, IN — Country Inn and Suites